Tourism, Memorials and Landscapes of Violence

The book focuses on tourism, memorial sites of the Holocaust and the Pacific War and the management practices for the visitors that they attract.

It provides an account of landscapes of violence as millions of people in Central and Eastern Europe, China, Japan and the United States were affected by wars, conflicts and crises. A special feature of the book is to reconstruct the changing management practices and the significance these heritage sites have attained for different visitor groups and the local populations, and to critically assess the current situation 80 years after the events. The book discusses the new directions of dark tourism, thanatourism and dissonance in heritage tourism in contemporary tourism research. Several case studies and in-depth analysis of memorial sites allow the reader to understand the consequences of past or ongoing policy changes.

This book will appeal to students and researchers in the fields of tourism, heritage, history, cultural studies, anthropology and human geography.

Rudi Hartmann is a Professor Emeritus (C/T) at the University of Colorado Denver, where he has taught geography and tourism planning since 1992. He received his Ph.D. in Geography from the Technical University of Munich, Germany, in 1983. A long-time interest of his is the study of tourist experiences at heritage sites. He has closely examined heritage tourism at memorial sites of the Holocaust in Germany and in the Netherlands. He has published numerous articles and books on these and related topics.

Routledge Cultural Heritage and Tourism Series

Series editor: Dallen J. Timothy

Arizona State University, USA

The Routledge Cultural Heritage and Tourism Series offers an interdisciplinary social science forum for original, innovative and cutting-edge research about all aspects of cultural heritage-based tourism. This series encourages new and theoretical perspectives and showcases ground-breaking work that reflects the dynamism and vibrancy of heritage, tourism and cultural studies. It aims to foster discussions about both tangible and intangible heritages, and all of their management, conservation, interpretation, political, conflict, consumption and identity challenges, opportunities and implications. This series interprets heritage broadly and caters to the needs of upper-level students, academic researchers, and policy makers.

Cultural Tourism and Cantonese Opera
Jian Ming Luo

Screen Tourism and Affective Landscapes
The Real, the Virtual, and the Cinematic
Edited by Erik Champion, Jane Stadler, Christina Lee and Robert Moses Peaslee

Cultural Heritage and Tourism in Africa
Edited by Dallen J. Timothy

Adaptive Reuse in Latin America
Cultural Identity, Values and Memory
Edited by José Bernardi

Tourism, Memorials and Landscapes of Violence
Remembering the Holocaust and the Pacific War
Edited by Rudi Hartmann

For more information about this series, please visit: https://www.routledge.com/Routledge-Cultural-Heritage-and-Tourism-Series/book-series/RCHT

Tourism, Memorials and Landscapes of Violence

Remembering the Holocaust
and the Pacific War

Edited by Rudi Hartmann

Routledge
Taylor & Francis Group

LONDON AND NEW YORK

First published 2025
by Routledge
4 Park Square, Milton Park, Abingdon, Oxon OX14 4RN

and by Routledge
605 Third Avenue, New York, NY 10158

Routledge is an imprint of the Taylor & Francis Group, an informa business

British Library Cataloguing-in-Publication Data
A catalogue record for this book is available from the British Library

Library of Congress Cataloging-in-Publication Data
Names: Hartmann, Rudi, editor.
Title: Tourism, memorials and landscapes of violence : remembering the
   Holocaust and the Pacific War / edited by Rudi Hartmann.
Description: New York : Routledge, 2025. | Series: Routledge studies in
   cultural heritage and tourism series | Includes bibliographical references
   and index.
Identifiers: LCCN 2024025891 (print) | LCCN 2024025892 (ebook) |
   ISBN 9780367423582 (hbk) | ISBN 9781032852157 (pbk) |
   ISBN 9780367823795 (ebk)
Subjects: LCSH: Dark tourism. | Heritage tourism—Management. |
   Holocaust, Jewish (1939–1945) | World War, 1939–1945—Campaigns—
   Pacific Ocean.
Classification: LCC G156.5.D37 T67 2025 (print) | LCC G156.5.D37
   (ebook) | DDC 940.53/1864—dc23/eng/20240712
LC record available at https://lccn.loc.gov/2024025891
LC ebook record available at https://lccn.loc.gov/2024025892

ISBN: 978-0-367-42358-2 (hbk)
ISBN: 978-1-032-85215-7 (pbk)
ISBN: 978-0-367-82379-5 (ebk)

DOI: 10.4324/9780367823795

Typeset in Times New Roman
by Apex CoVantage, LLC

Contents

Figures

Contributors

Hamilton Bean, PhD, MBA, APR, is Professor in the Department of Communication at the University of Colorado Denver. He also serves as Director of the University of Colorado Denver's International Studies Program. He specializes in the study of communication and security. His latest co-edited volume is *The Handbook of Communication and Security* (Routledge, 2019).

Bonnie J. Clark serves as Professor of Anthropology and Curator for Archaeology at the University of Denver (DU). Dr. Clark leads the DU Amache Project, a community collaboration committed to researching, preserving and interpreting the physical history of Amache, Colorado's WWII-era Japanese American incarceration camp. She is the author of numerous works, including *Finding Solace in the Soil: An Archaeology of Gardens and Gardeners at Amache*.

Manfred Deiler trained as a social security employee and worked as a certified social security clerk. In the late 1980s, he got involved in voluntary service regarding the history of the Dachau subcamp cluster at Kaufering and Landsberg, Germany. In 2009, he became the president of the newly founded European Holocaust Memorial Foundation in Landsberg and committed himself to the preservation of the last remnants of the camps and the creation of a memorial site.

Rudi Hartmann is Professor Emeritus (C/T) at the University of Colorado Denver, where he has taught geography and tourism planning since 1992. He received his PhD in Geography from the Technical University of Munich, Germany, in 1983. A long-time interest of his is the study of tourist experiences at heritage sites. He has closely examined heritage tourism at memorial sites of the Holocaust in Germany and in the Netherlands. He has published numerous articles and books on these and related topics.

Kyungjae Jang is Associate Professor at the Graduate School of Humanities and Social Sciences, Hiroshima University. He holds a PhD and MA in tourism studies from Hokkaido University and a BA in Korean History from Korea University. Dr. Jang is currently conducting research in military port cities in East Asia to understand how war memories are interwoven with popular culture to inherit memory and how it is represented in the form of tourism.

Whitney J. Peterson received a master's degree in anthropology from the University of Denver (DU) in 2018, where she participated in the DU Amache Project. For the last decade, Peterson has worked to preserve and amplify WWII history of Japanese-American incarceration having previously worked at the Manzanar National Historic Site from 2011 to 2014. Today, Peterson leads digital documentation projects at the non-profit organization, CyArk, where she works with communities and heritage managers to create place-based virtual experiences that promote connection with cultural heritage.

Edith Raim studied history and German literature at Munich university and spent a year as a visiting graduate student in Princeton, USA. From 1991 to 1995, she worked as a lecturer for the German Academic Exchange Service in the German Department of the University of Durham, later at the museum Haus der Geschichte der Bundesrepublik Deutschland in Bonn, the Institute of Contemporary History in Munich and Augsburg University. She publishes widely on 20th-century German history.

Richard Sharpley is Emeritus Professor of Tourism at the University of Central Lancashire, UK. His principal research interests are in the fields of tourism and sustainable development, the tourist experience and dark tourism.

Dietrich Soyez is Professor Emeritus of Geography at the University of Cologne, Germany. He was Visiting Professor with universities in China, Canada, France and Oman. He had a leading role at the IGU (International Geographical Union) as Vice President (2008 to 2012) and First Vice President (2012 to 2016). His principal research interests are, as a rule with a special focus on transnational implications: environmental economic geography (EEG), political geography shaped by civil society actors, industrial tourism and dark industrial heritage.

Ming Ming Su is Associate Professor at the School of Environment and Natural Resources, Renmin University of China, Beijing. She holds degrees from the University of Waterloo in Canada and Tsinghua University of China. Her research focuses on heritage management, tourism impacts, tourism and community relations, tourism at protected areas and tourism issues in China.

Preface and acknowledgments

The book focuses on tourism to memorial sites of the Holocaust and the Pacific War and the management practices for the visitors that they attract. It provides an account of landscapes of violence as millions of people in Central and Eastern Europe, China, Japan and the United States were affected by wars, conflicts and crises.

The major themes of the book came out of my initial research on tourism to memorial sites of the Holocaust in the 1970s and 1980s. My interest widened in two ways, conceptually with a review and discussion of several parallel terms used in the literature of the 1990s and 2000s including *dark tourism*, *thanatourism* and *dissonance* in heritage tourism studies (see Research Note from 2013) and regionally with the inclusion of sites of the Pacific War most notably Lu Gou Qiao (Marco Polo Bridge), marking the beginnings of the Second Sino-Japanese War in 1937 and Hiroshima bringing the fatal war events in Pacific-Asia to an end in 1945. The realization that most of the places have seen large and repeated commemorative events for the victims caused me to focus on visitation and tourism to the most prominent sites. While early research on Dachau – I am a native of Munich, Germany – and eventually at the Anne Frank House in Amsterdam and in Buchenwald in former East Germany exposed me to several now well-known, tragically infamous sites of the Holocaust, later research supported by my university, with trips to China and Japan, made me more familiar with the issues and problems associated with the sites of the Pacific War.

During my research efforts I found wide support and recognition from colleagues of the Association of American Geographers (AAG) and the International Geographical Union (IGU), in particular in several specialty groups (pertaining to the study of recreation and tourism and within the geography of China group). I am grateful for the continued professional encouragement I received locally, at the University of Colorado Denver, as well as in several countries internationally. Especially, I would like to thank Dallen Timothy, editor of the *Journal of Heritage Tourism*, for his support of my 2013 Research Note and other publications, and Phil Stone, co-founder of the Dark Tourism Forum at the University of Eastern Lancashire and principal editor of the 2018 *Palgrave Handbook of Dark Tourism Studies*.

My thanks also go to the editors and program managers within Routledge, a publication of Taylor and Francis, with headquarters in Oxford, UK. Early on, in

2018/2019, I was invited by Faye Leerink to contribute to their programs. I am especially grateful for her continued support during the years of the COVID pandemic when the development of the book project slowed down and at times came to a temporary stop. During the more recent book development phase in 2023 and 2024, the progress of the volume was in the good hands of Editorial Assistant Prachi Priyanka. I am thankful for her consistently constructive collaboration and guidance through the book production process.

Finally, I am grateful to my contributors on a rather complex book project, with mostly original chapters as well as reprinted texts to be included. I enjoyed working with my colleagues and have received many valuable hints regarding how to tackle some of the problems and how to complete the job. In many of their contributions, highly innovative materials are presented.

I would like to thank my family, in particular my wife, Kathy Newman, and my son, 'Lenzi' Lorenz Hartmann, for listening to my book issues over the years, their encouraging words and occasional help on several fronts.

The book is dedicated to the men and women who have provided valuable service at the memorial sites in honor of the victims of the Holocaust and the Pacific War. Frequently, the managers and educators at the sites had to communicate a difficult or even controversial heritage to the visiting public. After a long period of growth and expansion the memorial landscape has faced more recent challenges, most prominently in the form of a new historical revisionism playing out in several countries, including Germany, the Netherlands and Japan, in a highly divisive political climate. The site employees' efforts to maintain a precise and scientifically truthful narrative of the complex events should be lauded while at the same time a regular questioning of interpretations of the past is an appropriate and important practice to shield. At last, the work at the memorial sites has found in many places good recognition as well as was given the necessary financial support, and we should be defending these successes to date.

Rudi Hartmann
Denver, March 2024

Introduction

Main themes and structure of the book

Main themes: The present volume focuses on Memorials of the Holocaust and of the Pacific War. The years 1929–1945 were a tumultuous era marked by crises, war and genocide. The book examines the major political, economic and social conflicts that resulted in worldwide warfare and confrontations before peace was reached. Millions of lives perished or were affected in horrific ways. A large number of places in Europe and Pacific Asia continue to be closely associated with the tragic events. The book authors provide in-depth studies of the commemoration practices at selected historic sites with four and five case studies for Central and Eastern Europe and Pacific Asia, respectively. Some of the sites, most notably Dachau and Hiroshima, have become major destinations for international travelers as well as for educational tourism at home. The purpose of the volume is to reconstruct the changing management practices and the significance these heritage sites have attained for different visitor groups and the local populations and to critically assess the current situation 80 years after the events (1945–2024).

Structure of the book: the main ideas of the book are organized thematically, regionally and chronologically. *Thematically:* the book progresses from a general discussion of tourism to sites with a controversial history and the approaches used in the examination of such places to the presentation of nine case studies. Here, an analysis of the past and current situation at the memorial sites is given. *Regionally:* the book first introduces several memorial sites associated with the rise of Nazi Germany and the evolution of a European-wide concentration camp system, which resulted in the death of millions of people. Among the most systematically persecuted groups were Jews in Germany and the neighboring European countries, resulting in the genocide of European Jewry which has become known as the 'Holocaust' (or Shoah). After the discussion of memorials established in the honor of the victims of Nazi Germany, major sites associated with the Pacific War and the 'Pacific Theater' of World War II (WWII) are introduced. The complexity of the events and the eventual expansion of the Second Sino-Japanese War to a Pacific-wide War are shown for a number of memorial sites in China, the United States and Japan. *Chronologically:* The case studies featured in both regional parts are organized chronologically. Part I starts with a chapter on Dachau, the first Nazi Concentration Camp (1933). It would become the 'model' for 20 concentration camps developed by the Schutz Staffel (SS) in the following years. Each of the

DOI: 10.4324/9780367823795-1

main concentration camps had a larger number of subsidiary camps, in particular during the years of WWII. Dachau had ten satellite camps near the City of Landsberg in 1944/1945 (the 'Kaufering' camps), most of which were operating as labor camps for the future production of fighter planes. While the Dachau Memorial Site and Museum became a worldwide destination visited by millions, knowledge about the 'Kaufering' camps was suppressed and largely forgotten until a group of local historians revived research in support of the commemoration of the victims' lives. The chapter on the 'Kaufering' camps reconstructs the efforts to preserve the remaining sites and the development of a new documentation center. The following chapter focuses on the armament industry enabled by forced labor during WWII, as shown here in the aerospace sector. The development of new weapons and weapon systems would eventually have consequences for the post-WW future. The country that was longest occupied by Nazi Germany was the Netherlands (1940–1945). The chapter on how the City of Amsterdam lived through these years discusses various ongoing commemoration efforts, from the iconic secret annex where Anne Frank wrote her diary to the Verzets Museum, highlighting forms of resistance as well as the collaboration of the local population with the Nazi occupiers. Part II starts with a chapter on the conflict site Lu Gou Qiao near Beijing, where the Second Sino-Japanese War had its beginnings in 1937. Fifty years after the historic site, a memorial and museum of Chinese resistance against the Japanese Imperial Forces were established in 1987 – as an example and expression of red tourism and dark tourism in China. The Pacific-wide expansion of the war activities started with the Japanese surprise attack on Pearl Harbor (December 7, 1941) and the subsequent establishment of ten Japanese-American internment camps (1942–1945) in the American West. The chapter examines the commemoration of the events with the focus on the confinement of Japanese Americans in two internment camps, in Manzanar and Amache, by reconstructing the pilgrimages to both heritage sites and highlighting the memorial practice of photo albums. The following three chapters are concerned with different practices and trends in the commemoration of the war in Japan, with a discussion of the Yamato battleship narratives in contemporary Japanese popular culture, an analysis of 'Kamikaze' heritage tourism as a current expression in Japanese memorialization and the nuclear bombing of Hiroshima as commemorated at the Hiroshima Peace Park and experienced differently by victims and perpetrators as a case study shows.

1   Research note Dark tourism, thanatourism, and dissonance in heritage tourism management

New directions in contemporary tourism research (*Journal of Heritage Tourism*, 2014, Vol. 9, No. 2, 166–182)

Rudi Hartmann

This research note focuses on tourism to heritage sites with a controversial history and sites associated with death, disaster, and the macabre. Several new concepts and research directions have emerged in the study of such sites. Particular attention is given to the dark tourism and thanatourism approaches as well as to an analysis of dissonance in the management of heritage sites. Further, changes at places with a shadowed past are examined in the context of a revived geography of memory. There is a continued interest of the traveling public in revisiting war and peace memorials. In the final part of the research note examples of a new perspective on places of pain and shame are introduced.

Introduction

Over the past two decades, studies in heritage tourism have flourished and produced many results (e.g. Timothy, 2011). One the most intriguing trends has been a focus on heritage sites with a controversial history, including locations of war, atrocity and horror. These studies have fostered a debate over the nature of tourism at controversial sites. During the years 1995–2000, three new concepts were introduced into tourism studies: dissonance in heritage (tourism), thanatourism, and dark tourism. It was, in particular, dark tourism that caught on quickly and found resonance among many researchers and the media. Meanwhile multitudes of articles and books have been published, and several research initiatives have been launched.

This research note addresses several themes from the following perspectives:

- What initial publications gave rise to later studies in the field, and who presented these new conceptual directions?
- What reasons or fertile conditions germinated the new research foci and directions?
- What efforts to date ground and expand the research directions methodologically? How might they have influenced tourism studies?

DOI: 10.4324/9780367823795-2

- What other less tourism-centered approaches to the study of places with a shadowed history were pursued, in particular, in the field of historical and cultural geography with renewed efforts in a geography of memory?
- Finally, what attractions and lasting motivations have been reflected in heritage tourism before and after dark tourism entered the debate?

Initial publications 1995–2000

In the mid to late 1990s, three new terms appeared in the academic tourism literature denoting dissonance at contested heritage sites, including places of atrocity and tourists' fascination with death and tragedy: dissonant heritage, thanatourism, and dark tourism. In 1996, Tunbridge and Ashworth published their book *Dissonant heritage: The management of the past as a resource in conflict*. The same year, Seaton (1996) penned his seminal article, Guided by the dark: From thanatopsis to thanatourism. Foley and Lenon (1996) similarly coauthored a piece on visits to the Dallas memorial site of John F. Kennedy's tragic death. In 2000, Lennon and Foley re-introduced their initial research about tourism to the site of JFK's assassination and combined it with additional case studies in their book *Dark tourism: The attraction of death and disaster*, which helped disseminate and popularize the dark-tourism concept further.

Ashworth and Tunbridge (1990) devised the influential 'Tourist-Historic City' model in the subfield of urban tourism. Their detailed analysis of urban heritage eventually led them to develop the notion of heritage dissonance. They argue that dissonance is intrinsic to all forms of heritage – whatever the scale, context, or locale (Ashworth, 1994; Graham, Ashworth & Tunbridge, 2000; Tunbridge & Ashworth, 1996). Dissonance is implicit in commodification processes, in the creation of place products, and in the content of messages which may in some cases lead to disinheritance. Further, they discussed visitor motives and management strategies for atrocity sites, elaborating on how these motives and strategies differ between three groups: the victims, the perpetrators, and the (more or less uninvolved or innocent) bystanders. Tunbridge and Ashworth (1996) chose two examples for discussion: the Nazi concentration/extermination camps of Central and Eastern Europe, and sites pertaining to the heritage of apartheid in South Africa. In a separate publication, Ashworth (1996) examined the case of revived tourism at Krakow-Kazimierz, the former Jewish neighborhood in Krakow, which was featured in the 1993 movie *Schindler's List*.

The concept of thanatourism was introduced by Tony Seaton in the mid-1990s. In his 1996 article he recognized the deep fascination some visitors to battlefields and cemeteries have with death and dying, often identifying with those who are buried and remembered at these sites. His analysis explored the motives and lifeworld of thanatourists (motivated by the desire for actual or symbolic encounters with death), including present-day visitors to the battlefield site of Waterloo (Seaton, 1999) and those of World War I. In a later case study he reported his findings from wandering with British tourists through the large cemeteries of World War I (Seaton, 2002).

From this initial interest in tourists' fascination with death and dying emerged a much wider, albeit more nebulous, concept — dark tourism. Lennon and Foley

(2000), the initiators of this term, identified it as a distinct type of tourism but did not clearly define it. Their book describes dark-tourism behavior in a multitude of locales and destinations. And, they placed dark tourism into the wider context of modern tourism development:

> Our argument is that 'dark tourism' is an intimation of post-modernity. We do not seek to enter any philosophical debates over the use of this term but, rather, aim to recognize the significant aspects of 'post-modernity' which are broadly taken to represent its main features . . . . First, (critical features apparent in the phenomena) are that global communication technologies play a major part in creating the initial interest . . . Second, the objects of dark tourism themselves appear to introduce anxiety and doubt about the project of modernity . . . Third, the educative elements of sites are accompanied by elements of commodification . . .
>
> (Lennon & Foley, 2000, p. 11).

This assessment of dark tourism in contemporary society also separates Lennon and Foley's ideas from the work of Seaton, who maintains that expressions of thanatourism have been part of western civilization for a long time, with some of its roots going back to early practices in Christianity during medieval times. Seaton (1996, 2009) rejects the notion that forms of dark tourism and thanatourism essentially reveal a post-modern condition.

It is interesting to note that these tourism scholars shared a similar geographical and professional background in the 1990s. Seaton and Lennon and Foley, who are credited with pioneering similar ways of seeing tourism, were employed in southern Scotland. Foley (1995), coming initially from the cultural tourism studies field, left the research arena for administrative responsibilities by about 2000 (Seaton, personal communication (in e-mail), January 9, 2012). In the meantime, dark-tourism studies spread to other universities in England, Scotland, and Northern Ireland. A research center was recently established at the University of Central Lancashire, England, where a dark tourism forum website (www.dark-tourism.org.uk) has been hosted since September 2005.

Dark tourism is, according to Philip Stone, founder of the Dark Tourism Forum, 'the act of travel and visitation to sites, attractions and exhibitions which have real or recreated death, suffering or the seemingly macabre as a main theme' (Stone, 2005, n.p.). Thus, the innovation and idea center of the dark-tourism paradigm spread to central Lancashire, where the dark tourism research agenda expanded. There, dark tourism was given new conceptual dimensions and philosophical underpinnings, which will be discussed later in this essay.

Historical and disciplinary background

What may have been the reasons for the simultaneous emergence of the books, articles, and other studies noted above? At least two observations can be made. First, the general political climate changed dramatically in the years 1985–1995 and, among others things, allowed travel across the Iron Curtain. Second, the traditional

term 'cultural tourism' proved to be too narrow and confining, as it closely followed elitist forms of tourist behavior, precluding many expressions of popular
culture. Scholars, in particular in the UK and other parts of the English-speaking
world, preferred to use the broader concept of heritage instead.

It was likely no coincidence that Tunbridge and Ashworth (1996) and Lennon
and Foley's (2000) books both donned photographs of Auschwitz on their covers.
With the collapse of communism in Eastern Europe and the removal of travel barriers between the west and the east, many sites of Nazi atrocity in Poland, Eastern
Germany, Czechoslovakia, Hungary, and Ukraine became accessible. Further, the
sites of Gulag labor camps in the former Soviet Union, where millions of people
were subjected to tragedy as well, could be visited by western travelers (Shapiro,
1994). In South Africa, the Apartheid principle was abandoned with the first free
elections in 1994. The same year, the first exhibit about Robben Island, the detention camp for Nelson Mandela and others who opposed Apartheid, was shown in
Cape Town. In 1997, Robben Island, eventually a World Heritage Site, became
accessible to tourists (Tunbridge, 2005).

The political climate has changed, the Cold War alignment of states has broken,
and some observers have noted the role tourism had in these processes. It could
be argued that personal travel at the beginning of the breakup period of Eastern
Europe, such as the Pan-European Picnic on the border of Hungary and Austria,
which ushered in the collapse of the Iron Curtain, and 'freedom rides' from East
Germany through Hungary to Austria at that time, contributed to the massive political changes of 1989–1990 (O'Sullivan, 2012; Schweizer, 2000). Other scholars
have been more cautious or skeptical about the positive role of tourism in the peace
process and in the promotion of intercultural understanding (Ashworth, 2012a;
Tomljenovic & Faulkner, 2000).

One of the results of destinations opening throughout Eastern Europe was
the emergence of Auschwitz as a leading travel destination. By the late 1990s,
Auschwitz had surpassed Dachau as the most widely identifiable symbol of Nazi
atrocities (Marcuse, 2001, 2005). In the early 1990s, forms of dissonance at the
Auschwitz memorial site became evident. British geographer Charlesworth (1994)
examined communist and Catholic attempts to de-Judaize the place, which had
become for many in the Jewish Diaspora a sacred pilgrimage site. By the late 1990s,
Cole (1999) observed another phenomenon at Auschwitz: commercialization and a
commodification of the site now occasionally denoted as 'Auschwitz-land'.

'holocaust tourism' had evolved on a broad scale at hundreds of memorial sites
in Central and Eastern Europe (Ashworth, 1996, 2002; Tunbridge & Ashworth,
1996) with considerable help by researchers and government offices in Germany
and Austria. Some of the memorial sites in Germany have received consistently
half a million to a million visitors annually, such as the Berlin 'Topography of
Terror' exhibit on the former grounds of the Gestapo/SS headquarters in Berlin
(Nachama, 2012), the Dachau concentration camp memorial site near Munich and
the Buchenwald concentration camp near Weimar in the former East Germany.
The annual number of visitors to the Anne Frank House in Amsterdam, established
in 1960 at the Frank family hiding place, has surpassed one million since 2007

(Hartmann, 1989, 1997, 2003, 2004, 2005, 2012). Most visitors at these sites are high-school students and young adults.

The study of holocaust tourism emphasizes the meaning, value, and extent of visits to sites that honor victims of Nazi Germany. It can be considered a type of heritage tourism limited to the commemoration of lost lives and human tragedies in a distinct period (1933–1945). The main memorial sites and museums usually devote considerable attention to the sociopolitical processes leading up to the fatal empowerment of autocratic regimes with a racial ideology in Europe. For many visitors, participating in holocaust tourism is not so much about satisfying a curiosity to see a famous site associated with atrocities they might have seen on television or in a movie, but to learn in a more focused way about the losses of Jewish community life in Central and Eastern Europe and to commemorate the millions of lives lost in the Holocaust. In most places honoring the victims of Nazi Germany, holocaust tourism is a form of educational tourism. There are two agents with distinct goals in the educational processes at work at Holocaust sites: organizers of group visits (e.g. teachers carefully preparing an educational agenda for their audience, usually young people and students) and managers who try to facilitate a lasting learning experience (e.g. 'Dachau' as a 'Lernort', or place of learning and effective outdoor classroom).

Relatives and friends of the victims (or members of the same ethnic, social and/or political group) are also important visitors at these sites. Thus, an intended visit to Auschwitz represents a form of pilgrimage tourism (Cohen Ioannides & Ioannides, 2006).

With a widely established network of memorial sites honoring the victims of the German socialistic regime of 1933–1945, a secondary type of attraction has started to emerge in Germany and Austria: sites associated with Adolf Hitler and other leaders of the Third Reich. Thus, the 'Eagle's Nest'/Obersalzberg near Berchtesgaden, Hitler's second home and alternative government center, receives approximately 250,000–300,000 visitors annually. Subsequently, a new type of research has developed, which focuses on tourism to sites of victims and perpetrators. While Petermann (2012b) compared and contrasted tourism at Dachau and the Obersalzberg, John-Stucke (2012) examined visits to the Wewelsburg SS Nordic Academy and the adjoining Niederhagen Concentration Camp. The combined memorial site (since 2010) has been carefully developed and recently successfully integrated in a regional heritage tourism plan (Brebeck, 2008; John-Stucke, 2012).

While Nazi atrocity sites in Central and Eastern Europe became more easily accessible and eventually received increased visitation from abroad, it is noteworthy that outside the original sites, special museums were established to commemorate the Holocaust, most prominently in Jerusalem (Cohen, 2010; Krakover, 2005; Oren & Shani, 2012) and in Washington, DC, with the United States Holocaust Memorial Museum in 1993 (Cole, 1999). Dark-tourism researchers took note of the latter and included it in their research agenda. Again, Lennon and Foley (2000, p. 147, 155–6) pointed to the post-modern condition apparent in this new venture which has become, despite or because of an Americanization of the Holocaust events, another salient stage for dark tourism.

As noted above, besides political changes, conceptual changes in tourism studies and in the tourism nomenclature in the 1980s and 1990s helped open the door to dark tourism studies. The traditional (and limited) understanding of cultural attractions in tourism was substantially widened by the use of the much broader term 'heritage', which includes tangible and intangible elements of the past that are used for some purpose – here for tourist visits – in the present (Graham et al., 2000; Timothy, 2011; Timothy & Boyd, 2003). This conceptual preference began to show in the English tourism literature of the 1990s and was particularly evident in the scholarly discourse in the UK. In a prominent tourism geography textbook, for instance, Williams (2009, p. 248) introduced a far-reaching typology of heritage attractions that included not only the usual list of cultural sites but comprised all landscapes, builtscapes, workscapes, technoscapes, and peoplescapes.

The academic attention on heritage led to the establishment of two scientific journals in the field: the *International Journal of Heritage Studies* (established 1995) and the *Journal of Heritage Tourism* (established 2006). These have also become outlets for studies that have introduced dissonant heritage, thanatourism, and dark tourism into the vocabulary and research approaches of heritage tourism

In the German literature a comparable trend in covering *Kulturtourismus* was apparent (cf. Steinecke, 2007). The field now includes, for example, popular forms of *Industrietourismus* or visits to industrial heritage sites once considered awkward destinations within the elitist framework of cultural tourism (Lauterbach, 2012; Schroeder, 2003; Soyez, 1986). Similarly, much research has emerged in tourism studies to sites with a controversial history, revealing or highlighting the darker side of humanity. A recently published volume includes examples in this new evolving field from the German-speaking world (see Fasching 2012; Kueblboeck, 2012; Quack & Steinecke 2012; Wolf & Matzner 2012).

Progress in dissonant heritage and thanatourism/dark-tourism studies

The past decade saw manifold publications and research projects on heritage dissonance, thanatourism, and dark tourism (e.g. Biran & Poria, 2012; Graham & Howard, 2008; Stone, 2011; Timothy, 2011). In recent years, Ashworth and other heritage scholars focused increasingly on current issues and problems regarding the management of heritage sites. The inevitable contestation of heritage in multicultural societies – a hot political issue, particularly in Western Europe, has become a focus and extension of earlier work on heritage dissonance (Ashworth, 2002, 2008; Ashworth & Hartmann, 2005; Ashworth, Graham & Tunbridge, 2007; Poria & Ashworth, 2009). Ashworth, Graham and Tunbridge (2007) noticed a trend towards 'pluralizing pasts' in contemporary pluralistic societies, with more questions than answers for a constructive role of heritage tourism development at many sites. Ashworth (2012a, 2012b) raised critical questions about the role of heritage tourism in the aftermath of ethnic conflicts including whether or not tourism is part of the solution or part of the problem. Analyses of political conflicts in Ireland, Cyprus, Palestine, South Africa and Thailand have generated preliminary research results that confirm Ashworth's skeptical outlook on the perhaps overrated potential of heritage tourism. In some situations it may contribute positively, or negatively, to the resolution of ethnic or cultural divisions.

Ashworth (2008) continued to examine the implications of past human trauma and violence with regard to current expressions of heritage tourism, although he remained largely outside the debate over the death, disaster, and macabre-fueled motivations of tourists high-lighted by dark-tourism researchers. Ashworth (personal communication (in e-mail), March 8, 2009) argues that there are no dark sites, only dark tourists. It is also interesting to note that the early pioneering work of Tunbridge and Ashworth (1996) on dissonant heritage and tourism at the Nazi concentration camps in Central and Eastern Europe, was not recognized until the early 2000s by dark-tourism commentators.

Thanatourism saw continued interest among tourism scholars, particularly from the UK. Tony Seaton, the concept's initial proponent, became involved in joint work with Dann on thanatourism at slavery sites (Dann & Seaton, 2002). Dann (2005) eventually began examining the crossover between media and thanatourism. Dann grounded his studies in three humanistic subfields, wherein he combined a cultural anthropology of tourism, the linguistics of tourism, and a qualitative examination of media content. Dann argued that we have seen a convergence of media and tourism, with far-reaching implications for travel motivations and expectations. He shed light on what it means for young people growing up with current media fascination with violence and how this might be expressed in desires to visit dark attractions. Dann (2005) explained that tourists also show interest in the darker side of humanity and why some dark destinations are more popular than others. As society can no longer completely dissociate itself from violence, we all might be to some extent, according to Dann, children of the dark.

Seaton's (2009) case studies in England highlighted the nature of thanatourism, explaining how tourists' fascination with death and dying differs from the partially overlapping and parallel view of dark tourism. Newcomers to the subject have contributed substantial knowledge about thanatourism with studies from Bosnia (Johnston, 2011) and North Korea (Lee, Bendle, Yoon, & Kim, 2012). Most recently, Johnston (2011) introduced the term 'thanagaze' in reference to Urry's (1990) tourist gaze and in differentiating Stone's 'mortal gaze'. Miles (2012) recently applied the thanatourism and dark-tourism concepts to an examination of the meanings and interpretations of battlefield tourism.

In 2005, Philip Stone established a Dark Tourism Forum website, raising the status of dark tourism in the media and in tourism studies. Stone is an avid promoter of dark-tourism research and has made several conceptual contributions to the topic. He has also co-edited several influential volumes on contemporary tourism issues, frequently with Richard Sharpley (Sharpley & Stone, 2011, 2012; Stone & Sharpley, 2008, 2009).

In 2005, Stone noted the vagueness of the dark-tourism concept and a lack of theoretical grounding. Thus, a strategic alliance with researchers of thanatourism – the 'sister term' characterized by Stone (2006, p. 146) as 'awkward' but more precise – was forged. As well, methodological contributions in further grounding dark tourism were made. Among the conceptual extensions that Stone proposed was the formulation of a dark-tourism spectrum, with 'shades of grey' in darkness building on suggestions previously expressed by Miles (2002) and Sharpley (2005). Stone's detailed visualization of a dark-tourism spectrum refines the supply side (see Figure 1.1). The framework operates on two axes, one rating dark-tourism products from darkest to lightest, while the other details, for instance, the education

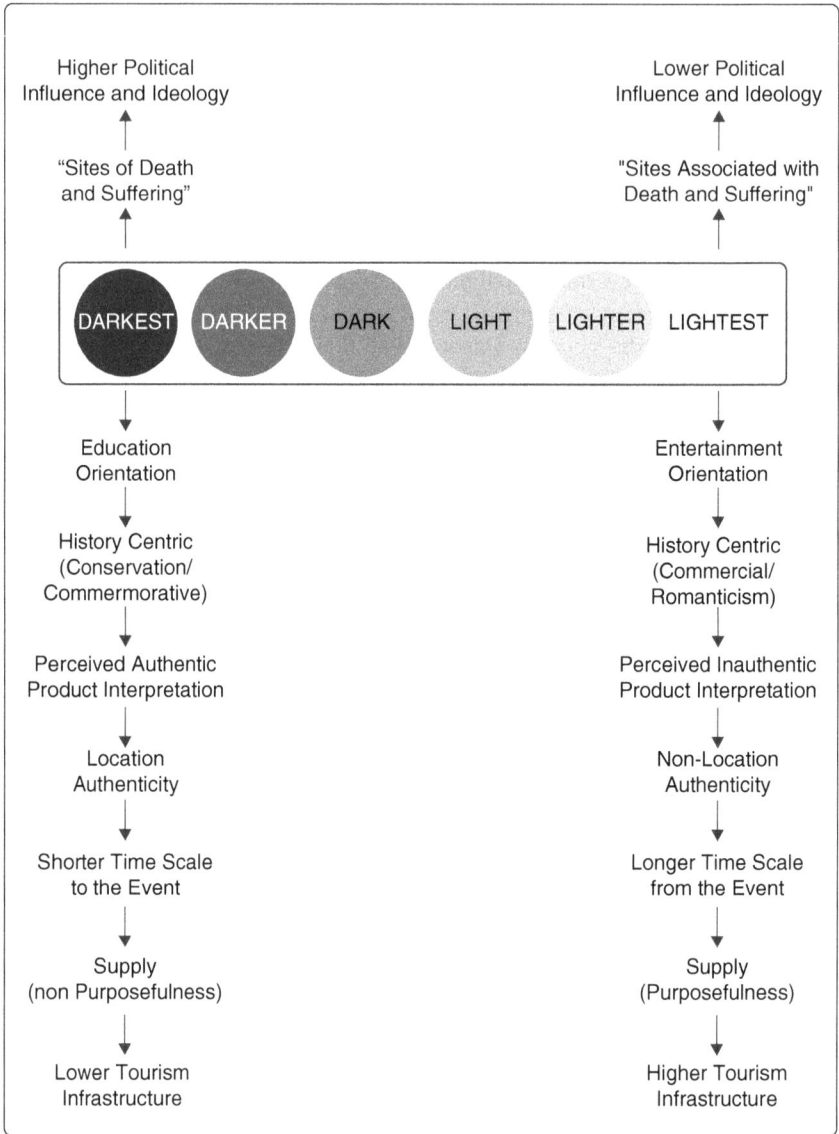

Higher Political Influence and Ideology	Lower Political Influence and Ideology
"Sites of Death and Suffering"	"Sites Associated with Death and Suffering"

DARKEST DARKER DARK LIGHT LIGHTER LIGHTEST

Education Orientation	Entertainment Orientation
History Centric (Conservation/ Commermorative)	History Centric (Commercial/ Romanticism)
Perceived Authentic Product Interpretation	Perceived Inauthentic Product Interpretation
Location Authenticity	Non-Location Authenticity
Shorter Time Scale to the Event	Longer Time Scale from the Event
Supply (non Purposefulness)	Supply (Purposefulness)
Lower Tourism Infrastructure	Higher Tourism Infrastructure

Figure 1.1 Dark Tourism spectrum (after Stone, 2006)

or entertainment orientation of a given site (e.g. a concentration camp memorial site versus a re-created Dracula Castle). With this framework, Stone (2006) outlined seven dark suppliers: dark fun factories, dark exhibitions, dark dungeons, dark resting places, dark shrines, dark conflict sites, and dark camps of genocide.

Following this conceptualization of dark-tourism supply, Stone and Sharpley (2008) defined the demand side conceptually. They waded into the troubled waters of a generally perceived public neglect of death in modern societies and a subsequent demise of traditional death rituals (Giddens, 1991). They found that

> in linking the concept of dark tourism with the sociology of death, the paper has not only developed a model that provides a conceptual basis for the further empirical study of its consumption, but has also contributed to wider social scientific understanding of mechanisms for confronting death in contemporary societies.
>
> (Stone & Sharpley, 2008, p. 589).

In short, the common absence of death in the public realm in western societies has resulted in specific forms of dark-tourism behavior. Recently, Stone (2012) expanded his exploration of a perceived "thanalogical condition of contemporary society'. His model focuses on dark-tourism experiences within a theoretical mediating mortality framework, which provides greater clarification of the demand for darkness. Stone analyzed the complex relationships between ethics and dark-tourism studies, and he pointed to a need for a new post-disciplinary research agenda (Stone, 2011).

The achievements in this field to date are remarkable, mostly regarding how dark tourism came to the fore of tourism studies. Stone's use of the Internet in attracting the media to dark- tourism themes was pioneering. As well, Stone, Sharpley, Seaton and others succeeded in bringing tourism themes once considered peripheral, such as the macabre, eclectic or non-traditional, into mainstream academia.

Critics of dark tourism and thanatourism became vocal as the new research agenda grew. More than a decade after Lennon and Foley's (2000) classic book was published, questions still abound, such as what is dark tourism or what is dark about dark tourism (Bowman & Pezzullo, 2010), what makes this approach different from older or competing conceptionalizations such as black spots tourism or morbid tourism (Blom, 2000; Rojek, 1993), and what might the different strands of dark tourism have in common. What do tourists' interest in Jack the Ripper walking tours in London, personal pilgrimages to Auschwitz, 'slumming' experiences in the favelas of Rio de Janeiro or at other shantytowns and 'urban jungles' (Frenzel & Koens, 2012; Frisch, 2012; Hartmann & Nagel, 2012; Rolfes, 2011; Steinbrink, 2012) have in common with the serene sites of war cemeteries (Vanneste, 2012)? Two larger sets of critiques are found in the works of Biran and Poria (2012) and Jamal and Lelo (2011). The latter emphasize that darkness is inherently a socially constructed concept.

The persisting cultural limitations of the term dark tourism show linguistically as researchers from non-English-speaking countries have preferred to use the English term.

'Dunkler Tourismus', the direct German translation, does not exude the 'emotive label' the way the term dark tourism does (Stone, 2006, p. 146). Neither has the French term 'le tourisme noir' caught on the way 'film noir' was successfully applied to a type of movie from the late 1940s. In other non-Western civilizations the term may not be translatable at all. It is questionable, for instance, whether the meaning of dark (as in the darker side of humanity) can be properly expressed within the polar opposites of yin and yang in Asian philosophy. The yin yang concepts are understood as natural contrasts, in black and white (not in shades of grey!) and, further, as complementary features that define the human condition.

From a history of tourism thought perspective, dark tourism has largely remained a British notion. In fact, most researchers and practitioners of dark tourism were based in southern Scotland and northern England. As dark-tourism proponents lived in relative proximity to each other and a dense regional network in the academy developed, it certainly helped integrate the dark-tourism subfield into broader tourism studies in Britain (Sharpley & Stone, 2011, 2012).

Dark tourism has come of age. A new Institute for Dark Tourism Research was inaugurated at the University of Central Lancashire in April 2012. The event reflects maturation and a further institutionalization of the topic. Also indicative of a maturing subject, several detailed histories have been published (Biran & Poria, 2012; Seaton, 2009; Stone, 2011).

One of the most interesting extensions of the dark-tourism paradigm comes from Israeli scholars Biran and Poria (2012) in their deconstruction and re-conceptualization of dark tourism. They argue that deviant behavior, as socially non-acceptable 'dark' behavior, is the true frontier in dark-tourism research. While there are acts of positive deviance (e.g. highly unconventional forms of behavior in far-away destinations), thoroughly improper acts cannot be communicated at home (negative deviance).

With few exceptions, tourism research has yet to address the deviant behavior of travelers as a common practice. One theme that has started addressing inappropriate forms of behavior is sex tourism. Qualitative studies have occasionally shed light on some topics of mixed deviance. At home, people rarely reveal deviant sexual behavior undertaken on holiday, whether it involved child prostitution or the potential spread of HIV by interacting with sex workers. On a different front, tourists rarely voice misgivings, remorse or shame about the environmental damage they contribute to by driving recklessly over sensitive high-elevation tundra biomes or by standing on coral reefs while diving. Are these examples proof of a truly dark side of travel and tourism, darker than the hidden or repressed fascination with death and disaster? In retrospect, European travel abroad brought about catastrophic and traumatic consequences, from fatal epidemics among native peoples in sixteenth century to mass killings of North American bison by big game hunters in the 1800s. Surely there are other approaches that look at the darker side of humanity and have made significant contributions to the study of places with a controversial history.

Geography of memory: a new research direction about places with a shadowed history

Considerable efforts have been made by geographers in the USA to revive a 'geography of memory' first outlined by Lowenthal (1985) in his book *The Past is a foreign country* and earlier writings. Other important contributions came from Foote (1997), who examined America's landscapes of violence and tragedy. His examples show how intense controversies arise over historic sites. Foote highlights four major outcomes for places associated with tragic events. The most common outcome is the process of rectification, which can lead to the designation of a site and eventually pave the way to sanctification. On the other end of his proposed continuum is the process of obliteration, a frequent situation in places that experienced acts of violence and tragedies now forgotten in time (Figure 1.2). Foote (2009, pp. 38–39) maintains:

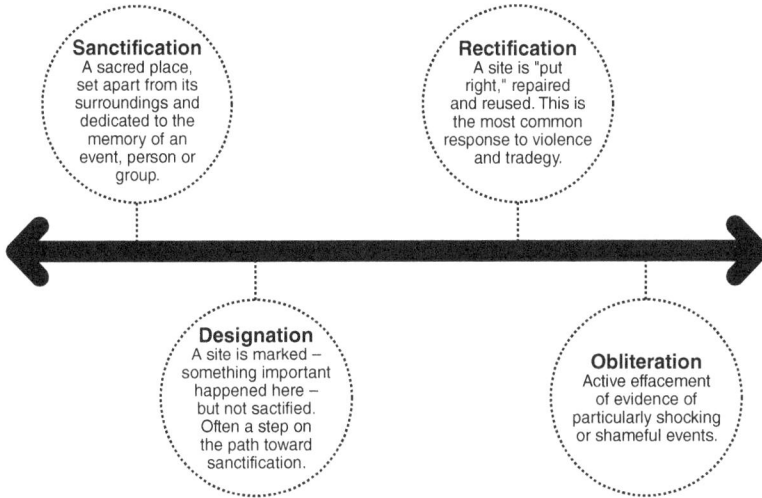

Figure 1.2 Common outcomes for heritage places associated with violence and tragedy (after Foote, 2009, p. 43).

. . . no one outcome is ever final. Sanctification, designation, rectification, and obliteration are not static outcomes, but only steps in a process. Almost all of the historical sites I have studied change through time, sometimes quite dramatically, and even outcomes like sanctification may take years, decades or centuries to occur. Sanctified or designated sites might be vandalized or destroyed to protest the value they embody. Obliterated sites may be rediscovered and marked, again as protest, but sometimes to acknowledge belatedly the victims of a long-past tragedy.

Interestingly, there seems to be no distinct set of rules about when and how a place with a shadowed past enters the process of designation and sanctification. It may take a few years, several decades or more than a century to designate a place finally as a public memorial site, which can and will be visited equally by insider and outsider groups related to a specific tragic event. Three examples from Colorado (USA) illustrate the seemingly arbitrary nature of site sanctification:

(1) Spontaneous shrines for the victims of the Columbine High School shootings immediately after the tragic events in Littleton, Colorado, in 1999, with a permanent memorial in a nearby park dedicated on 21 September 2007 – following intense debates over 13 erected wooden crosses including two for the perpetrators who committed suicide after the shooting rampage (Foote & Grider, 2010; Hartmann, 2009).
(2) The 2006 decision for the Camp Amache National Historic Landmark status, more than 60 years after 10,000 Japanese-Americans were relocated and forced to live in a large internment camp outside Granada, Colorado, 1942–1945 (Hartmann, 2009).
(3) The Sand Creek Massacre National Historic Site, which opened to the public near Chivington, Colorado, in 2007, over 140 years after Colonel John Chivington led an attack on a camp of peaceful Southern Cheyenne and Arapahoe Indians in 1864 (Hartmann, 2011).

Owen and Alderman (2008) applied the geography of memory approach to places associated with slavery and the civil rights movement in the southern USA. Both the persistent inhumane treatment and social conditions of African-Americans in the past and the later political processes in protest against them have resulting in many monuments and memorials in the struggle for civil rights in the 'deep South' (Alderman, 2006, 2009).

Recently the US National Park Service implemented new policies regarding the protec-tion of sites associated with slavery and the civil rights movement, massacres of Native American tribes (Timothy & Boyd, 2003), the internment of Japanese-Americans during World War II, and other 'Places of Conscience and Places of Commemoration', the title of a 2007 special issue of *Common Ground*, a magazine published by the National Park Service.

Travelers are reminded in guide books and websites of landmarks imbued with deep national pride that are often considered shrines of a nation (Timothy, 2011). In the USA, the homes of national founders certainly fall into this attraction category, and yet new research perspectives and discoveries at George Washington and Thomas Jefferson's homes have led to a partial re-evaluation of race relations in eighteenth- and nineteenth-century America. The complex relations the Washington family and Thomas Jefferson personally had with slaves are now under scrutiny and major points of debate (Casper, 2008; Gordon-Reed, 2008; Stanton, 2002). Again, the geography of memory helps reflect on the processes of sanctification at emotive national landmarks.

Conclusion: a continued interest in revisiting war and peace memorials

Arguably, dark attractions make up only a portion of the contemporary tourism spectrum (Biran & Poria, 2012; Sharpley, 2005) among a multitude of heritage attractions and destinations (Timothy, 2011; Timothy & Boyd, 2003; Williams, 2009). Tourism landscapes and monuments associated with war, military conflict, and peace continue to attract large numbers of visitors. In her examination of the relationships between war and tourism, Smith (1998, p. 202) argued that 'war-related tourism attractions are the largest single category known'. Most recently, Butler and Suntikul (2012) reiterated that war-related heritage has become a highly valued tourism commodity. It is widely shared among the traveling public whether the actions that led to war, conflict, and eventually peace are supported, approved of in an apologetic and forgiving way, or condemned outright. Long before dark-tourism research began, memorials to wars won or lost and in honor of their victims dotted the cultural landscape.

There is a strong interest in the landscapes left by both world wars in twentieth-century Europe. An example of how the war sites and memorial landscapes have gradually and consistently changed with the introduction of memorial events and the provision of tourism services are the World War I battlefield sites in Flanders, Belgium. As the centenary approaches (1914–2014), Jansen-Verbeke and George (2012) distinguish several distinct stages that have occurred over the past century: from war landscapes to memoryscapes, from memoryscapes to heritage landscapes, and from heritage landscapes to tourism landscapes. Vanneste and Foote (2009) examined the preparations made on local and regional levels for the centennial in Flanders Field and asked the question of whether the efforts will aim to support and expand the existing war tourism sector or to reorient the meaning of the battlefields toward peace and reconciliation. The latter concern is the point of departure for Petermann (2007, 2012a) in her examination of the changing rituals at the memorial landscapes in Verdun, France.

In the USA, the commemoration of military conflicts is frequently shared in the form of a living history of events and by reenacting battles. Reenactments have become a popular movement among history buffs where, for instance, the battles of the War of Independence are replayed in great detail. As the northeastern and southern states of the USA celebrate the 150th anniversary of the Civil War, battle reenactments involved many participants. At reenacted Civil War battles there are no longer winners and losers. Instead, a sense of camaraderie prevails among the actors, who are now more concerned about wearing historically correct period uniforms or accurately portraying the figures of yesteryear (Hadden, 1999). Some major reenacted battles, such as Manassas and Gettysburg, have attracted tens of thousands of participants and spectators in recent years. The National Park Service does not allow battle reenactments on its properties where combat is simulated; only the living history approach is permitted (e.g. the lifestyle of average Civil War soldiers) (Marquis, 2008; Strauss, 2001).

In many places and cultures, wars and military conflicts have seen a more or less playful reenactment of selected events as well. Recently, the 90th anniversary of the foundation of the Chinese Communist Party (CCP) saw many popular reenactments. Thousands of young party members joined in the reenactment of the Long March in different provinces of China and dressed up in historic uniforms. The CCP also decided to use reenactments to showcase the 'liberation' of Tibet 50 years after the 1950 military invasion in a public film production on site (Simons & McCurry, 2002).

In Germany, local traditions have given rise to a multitude of dark historical reenact-ments, from the replay of Rothenburg's legendary 'Master Draught' that saved the city in the seventeenth-century Thirty Years' War, to the solemn silent walk (Erinnerungsgang) remembering the 42 Jewish victims of the 1938 Kristallnacht pogrom in Oldenburg, an annual event since 1981 (www.erin nerungsgang.com).

Logan and Reeves' (2009) book, *Places of Pain and Shame*, introduced the term 'difficult heritage' into the dark-tourism lexicon. The book continues the dark-tourism discourse, including chapters on massacre and genocide sites, wartime internment sites, civil and political prisons, and places of benevolent internment. The nomenclature and approaches used in the book are appropriate in dealing with the difficult heritage of re-occurring wars and peace processes without discussing or even mentioning dark-tourism behavior at the chosen heritage sites, with one exception: Young's (2009) chapter on Auschwitz-Birkenau. The book is foremost about the universal sense of shame and pain humankind feels, or should feel, about past injustices and human-induced atrocities. As this new volume and other publications show, there is life in heritage tourism studies outside the dark-tourism/thanatourism agendas and that we have entered an era of globalization in the wider tourism studies arena, surprisingly with more voices joining in from areas outside Europe and North America.

References

Alderman, D. (2006). Naming streets after Martin Luther King, Jr.: No easy road. In R. Schein (Ed.), *Landscape and race in the United States* (pp. 213–236). Routledge.

Alderman, D. (2009). Writing on the Graceland wall: On the importance of authorship in pilgrimage landscapes. In O. Johansson & T. Bell (Eds.), *Sound, society, and the geography of popular music* (pp. 53–65). Ashgate.

Ashworth, G. J. (1994). From history to heritage, from heritage to identity: In search of concepts and models. In G. J. Ashworth & P. J. Larkham (Eds.), *Building a new heritage: Tourism, culture and identity in the new Europe* (pp. 13–30). Routledge.

Ashworth, G. J. (1996). Holocaust tourism and Jewish culture: The lessons of Krakow-Kazimierz. In M. Robinson, N. Evans, & P. Callaghan (Eds.), *Tourism and cultural change* (pp. 1–12). Centre for Travel and Tourism.

Ashworth, G. J. (2002). Holocaust tourism: The experience of Krakow-Kazimierz. *International Research in Geographical and Environmental Education, 11*(4), 363–367.

Ashworth, G. J. (2008). The memorialisation of violence and tragedy: Human trauma as heritage. In B. Graham & P. Howard (Eds.), *The Ashgate companion to heritage and identity* (pp. 231–244). Ashgate.

Ashworth, G. J. (2012a, February). *Ethnic conflict: Is heritage tourism part of the solution or part of the problem?* Paper presented at the annual meeting of the Association of American Geographers, New York, NY.

Ashworth, G. J. (2012b). Do tourists destroy the heritage they have come to experience? In T. V. Singh (Ed.), *Critical debates in tourism* (pp. 278–286). Channel View.

Ashworth, G. J., Graham, B., & Tunbridge, J. E. (2007). *Pluralising pasts: Heritage, identity and place in multicultural societies*. Pluto.

Ashworth, G. J., & Hartmann, R. (2005). *Horror and human tragedy revisited: The management of sites of atrocities for tourism*. Cognizant.

Ashworth, G. J., & Tunbridge, J. E. (1990). *The tourist-historic city*. Belhaven.

Biran, A., & Poria, Y. (2012). Re-conceptualizing dark tourism. In R. Sharpley & P. Stone (Eds.), *The contemporary tourist experience: Concepts and consequences* (pp. 62–79). Routledge.

Blom, T. (2000). Morbid tourism: A postmodern market niche with an example from Althorpe. *Norwegian Journal of Geography, 54*(1), 29–36.

Bowman, M. S., & Pezzullo, P. C. (2010). What's so 'dark' about 'dark tourism'?: Death, tours, and preference. *Tourist Studies, 9*(3), 187–202.

Brebeck, W. E. (2008). Wewelsburg 1933–45: Ansaetze und Perspektiven zur Neukonzeption deer Dauerausstellung. In W. E. Brebeck & B. Stambolis (Eds.), *Erinnerungsarbeit kontra Verklaerung der NS-Zeit. Vom Umgang mit Tatorten, Gedenkorten und Kultorten* (pp. 119–141). Verlag Dr. Christian Mueller-Straten.

Butler, R., & Suntikul, W. (Eds.), (2012). *Tourism and war: A complex relationship*. Routledge.

Casper, S. (2008). *Sarah Johnson's Mount Vernon: The forgotten history of an American shrine*. Hill and Wang.

Charlesworth, A. (1994). Contesting places of memory: The case of Auschwitz. *Environment and Planning D: Society and Space, 12*, 579–593.

Cohen, E. H. (2010). Educational dark tourism at an in populo site: The Holocaust Museum in Jerusalem. *Annals of Tourism Research, 38*(1), 193–209.

Cohen Ioannides, M. W., & Ioannides, D. (2006). Global Jewish tourism: Pilgrimages and remembrance. In D. J. Timothy & D. H. Olsen (Eds.), *Tourism, religion and spiritual journeys* (pp. 156–171). Routledge.

Cole, T. (1999). *Selling the Holocaust: From Auschwitz to Schindler — how history is bought, packaged and sold*. Routledge.

Dann, G. (2005). Children of the dark. In G. J. Ashworth & R. Hartmann (Eds.), *Horror and human tragedy revisited: The management of sites of atrocities for tourism* (pp. 233–252). Cognizant.

Dann, G., & Seaton, A. V. (2002). *Slavery, contested heritage and thanatourism*. Taylor & Francis.

Fasching, G. (2012). Erinnerungstourismus in Oesterreich: Die gegenwaertigen Ansaetze zur Erweiterung des Tourismusangebotes, zur Bewahrung des kulturellen Erbes and zur Staerkung des laendlichen Raums. In H. Quack & Steinecke, A. (Eds.), *Dark Tourism: Faszination des Schreckens* (pp. 23–46). University of Paderborn Press.

Foley, M. (1995). Cultural tourism in the United Kingdom. In G. Richards (Ed.), *Cultural tourism in Europe* (pp. 204–224). CABI.

Foley, M., & Lenon, J. (1996). JFK and dark tourism: A fascination with assassination. *International Journal of Heritage Studies, 2*(4), 198–211.

Foote, K. (1997). *Shadowed ground: America's landscapes of violence and tragedy*. University of Texas Press.

Foote, K. (2009). Heritage tourism, the geography of memory, and the politics of place in Southeastern Colorado. In R. Hartmann (Ed.), *The Southeast Colorado heritage tourism project report* (pp. 37–50). Wash Park Media.

Foote, K., & Grider, S. (2010). MemorialisSation of US college and universities tragedies: Spaces of mourning and remembrance. In A. Maddrell & J. D. Sidaway (Eds.), *Deathscapes* (pp. 181–206). Ashgate.

Frenzel, F., & Koens, K. (2012). Slum tourism: Developments in a young field of interdisci-plinary tourism research. *Tourism Geographies, 14*(2), 195–212.

Frisch, T. (2012). Glimpses of another world: The favela as a tourist attraction. *Tourism Geographies, 14*(2), 320–338.

Giddens, A. (1991). *Modernity and self identity*. Polity.

Gordon-Reed, A. (2008). *The Hemingses of Monticello: An American family*. W.W. Norton.

Graham, B., Ashworth, G. J., & Tunbridge, J. E. (2000). *A geography of heritage*. Arnold.

Graham, B., & Howard, P. (2008). *The Ashgate companion to heritage and identity*. Aldershot.

Hadden, R. L. (1999). *Reliving the civil war: A reenactor's handbook*. Stackpole Books.

Hartmann, R. (1989). Dachau revisited: Tourism to the memorial site and museum of the former concentration camp. *Tourism Recreation Research, 14*(1), 41–47.

Hartmann, R. (1997). Dealing with Dachau in geographic education. In H. Brodsky (Ed.), *Visions of land and community: Geography in Jewish studies* (pp. 357–369). University of Maryland Press.

Hartmann, R. (2003). Zielorte des Holocaust Tourismus im Wandel: die KZ-Gedenkstaette in Dachau, die Gedenkstaette in Weimar-Buchenwald und das Anne-Frank-Haus in Amsterdam. In C. Becker, H. Hopfinger, & A. Steinecke (Eds.), *Handbuch der Geographie der Freizeit und des Tourismus* (pp. 297–308). Oldenburg.

Hartmann, R. (2004). Das Anne-Frank-Haus in Amsterdam: Lernort, Literarische Land-schaft und Gedenkstaette. In A. Brittner-Widmann Quack & H. Wachowiak (Eds.), *Festschrift C. Becker* (pp. 131–142). University of Trier Press.

Hartmann, R. (2005). Holocaust memorials without Holocaust survivors: The management of museums and memorials to victims of Nazi Germany in 21st century Europe. In G. J. Ashworth & R. Hartmann (Eds.), *Horror and human tragedy revisited: The management of sites of atrocities for tourism* (pp. 89–107). Cognizant.

Hartmann, R. (Ed.). (2009). *The Southeast Colorado heritage tourism project report*. Wash Park Media.

Hartmann, R. (2011). From living history at Bent's Old Fort along the Historic Santa Fe Trail (1833–49) to the Commemoration of Death and Disaster at the Sand Creek Massacre (1864): Regional and thematic connectedness of two heritage sites in Southeast Colo-rado. In A. Kagermeier & T. Reeh (Eds.), *Studien zur Freizeit- und Tourismusforschung* (pp. 85–102). University of Trier Press.

Hartmann, R. (2012, February). *Tourism management and the role of the new media and arts at the Anne Frank House in Amsterdam: A museum and literary landscape goes virtual reality*. Paper presented at the Annual Meeting of the Association of American Geogra-phers, New York City, NY.

Hartmann, R., & Nagel, V. (2012). Township-Tourismus in Suedafrika. In H. Quack & A. Steinecke (Eds.), *Dark tourism: Faszination des Schreckens* (pp. 277–290). University of Paderborn Press.

Jamal, T., & Lelo, L. (2011). Exploring the conceptual and analytical framing of dark tour-ism: From darkness to intentionality. In R. Sharpley & P. Stone (Eds.), *Tourist experience: Contemporary perspectives* (pp. 29–42). Routledge.

Jansen-Verbeke, M., & George, W. (2012). Reflections on the great war centenary: From warscapes to memoryscapes in 100 years. In R. Butler & W. Suntikul (Eds.), *War and tourism: A complex relationship* (pp. 273–287). Routledge.

Johnston, T. (2011). Thanatourism and the commodification of space in post-war Croatia and Bosnia. In R. Sharpley & P. Stone (Eds.), *Tourist experience: Contemporary perspec-tives* (pp. 43–56). Routledge.

John-Stucke, K. (2012). Die Wewelsburg: Renaissanceschloss – "SS-Schule" – Erinnerung-sort – Ausflugsziel" In H. Quack & A. Steinecke (Eds.), *Dark tourism: Faszination des Schreckens* (pp. 179–192). University of Paderborn Press.

Krakover, S. (2005). Attitudes of Israeli visitors towards the holocaust remembrance site of Yad Vashem. In G. J. Ashworth & Hartmann, R. (Eds.), *Horror and human tragedy revisited: The management of sites of atrocities for* tourism (pp. 108–117). Cognizant.

Kueblboeck, S. (2012). Sich selbst an dunklen Orten begegnen: Existenzielle Authenzitaet als Potenzial des Dark Tourism. In H. Quack & A. Steinecke (Eds.), *Dark Tourism: Faszination des Schreckens* (pp. 113–126). University of Paderborn Press.

Lauterbach, B. (2012). Blut, Schweiss und Traenen: die dunklen Seiten des Industrietourismus. In H. Quack & A. Steinecke (Eds.), *Dark Tourism: Faszination des Schreckens* (pp. 102–112). University of Paderborn Press.

Lee, C.-K., Bendle, L. J., Yoon, Y.-S., & Kim, M.-J. (2012). Thanatourism or peace tourism: Perceived value at a North Korean resort from an indigenous perspective. *International Journal of Tourism Research, 14*(1), 71–90.

Lennon, J., & Foley, M. (2000) *Dark tourism: The attraction of death and disaster*. Continuum.

Logan, W., & Reeves, K. (Eds.). (2009). *Places of pain and shame: Dealing with 'difficult heritage*. Routledge.

Lowenthal, D. (1985). *The past is a foreign country*. Cambridge University Press.

Marcuse, H. (2001). *Legacies of Dachau: The uses and abuses of a concentration camp, 1933–2001*. Cambridge University Press.

Marcuse, H. (2005). Reshaping Dachau for visitors: 1933–2000. In G. J. Ashworth & R. Hartmann (Eds.), *Horror and human tragedies revisited: The management of sites of atrocities for tourism* (pp. 118–148). Cognizant.

Marquis, C. (2008). A history of history: The origins of war re-enacting in America. *McNair Scholars Research Journal, 1*(1), 1–16.

Miles, S. (2012). War memorials on the Western Front: British tourists and the embodiment of memory. In A. Kagermeier & J. Saarinen (Eds.), *Transforming and managing destinations: Tourism and leisure in a time of global change and risk* (pp. 179–186). University of Trier Press.

Miles, W. (2002). Auschwitz: Museum interpretation and darker tourism. *Annals of Tourism Research, 29*(4), 1175–1178.

Nachama, A. (2012). Die fuerchterlichste Adresse in Berlin: zur Konzeption eines Lernortes auf dem Gelaende der Gestapo, SS und des Reichssicherheitshauptamtes. In H. Quack & A. Steinecke (Eds.), *Dark tourism: Faszination des Schreckens* (pp. 153–170). University of Paderborn Press.

Oren, G., & Shani, A. (2012). The Yad Vashem Holocaust Museum: Educational dark tourism in a futuristic form. *Journal of Heritage Tourism, 7*(3), 255–270.

O'Sullivan, J. (2012). Hungary and the Cold War. *Hungarian Review, 5*, 57–64.

Owen, D., & Alderman, D. (2008). *Civil rights memorials and the geography of memory*. University of Georgia Press.

Petermann, S. (2007). *Rituale machen Raeume: Zum kollektiven Gedenken der Schlacht von Verdun und der Landung in der Normandie*. Transcript.

Petermann, S. (2012a, August). *From triumph to reconciliation: Rituals and tourism in Verdun*. Paper presented at the IGU Pre-conference "Transforming and managing destinations: Tourism and leisure in a time of global change and risks", Trier.

Petermann, S. (2012b). You get out of it what you put into: Nationalsozialistische Opfer- und Taeterorte in Deutschland als Touristenorte. In H. Quack & A. Steinecke (Eds.), *Dark tourism: Faszination des Schreckens* (pp. 63–80). University of Paderborn Press.

Poria, Y., & Ashworth, G. J. (2009). Heritage tourism: Current resource for conflict. *Annals of Tourism Research, 36*(3), 522–525.

Quack, H., & Steinecke, A. (Eds.), (2012). *Dark Tourism: Faszination des Schreckens*, Paderborner Geographische Studien zu Tourismusforschung und Destinationsmanagement. University of Paderborn Press.

Rojek, C. (1993). *Ways of escape: Modern transformations in leisure and travel*. Macmillan.

Rolfes, M. (2011). Slumming: Empirical results and observational-theoretical considerations on the backgrounds of township, favela and slum tourism. In R. Sharpley & P. Stone (Eds.), *Tourist experience: Contemporary perspectives* (pp. 59–75). Routledge.

Schroeder, A. (2003). Industrietourismus. In C. Becker, H. Hopfinger, & A. Steinecke (Eds.), *Geographie der Freizeit und des Tourismus* (pp. 213–224). Oldenbourg.

Schweizer, P. (2000). *The fall of the Berlin wall: Reassessing the cases and consequences of the end of the Cold War*. Hoover Institution Press.

Seaton, A. V. (1996). Guided by the dark: From thanatopsis to thanatourism. *International Journal of Heritage Studies, 2*(4), 234–244.

Seaton, A. V. (1999). War and thanatourism: Waterloo 1815–1914. *Annals of Tourism Research, 26*(1), 130–158.

Seaton, A. V. (2002). Another weekend away looking for dead bodies: Battlefield tourism on the Somme and in Flanders. *Tourism Recreation Research, 25*(3), 63–78.

Seaton, A. V. (2009). Thanatourism and its discontents: An appraisal of a decade's work with some future issues and directions. In T. Jamal & M. Robinson (Eds.), *The Sage handbook of tourism studies* (pp. 521–542). Sage.

Shapiro, M. (1994, September 4). The first Gulag. *Washington Post*.

Sharpley, R. (2005). Travels to the edge of darkness: Towards a typology of "dark tourism". In C. Ryan, S. Page, & M. Aicken (Eds.), *Taking tourism to the limits: Issues, concepts and managerial perspectives* (pp. 187–198). Elsevier.

Sharpley, R., & Stone, P. (2011). *Tourist experience: Contemporary perspectives*. Routledge.

Sharpley, R., & Stone, P. (2012). *Contemporary tourist experiences: Concepts and consequences*. Routledge.

Simons, L., & McCurry, S. (2002, April). Tibetans: Moving forward, holding on. *National Geographic Magazine*, 2–37.

Smith, V. (1998). War and tourism: An American ethnography. *Annals of Tourism Research, 25*(1), 202–227.

Soyez, D. (1986). Industrietourismus. *Erdkunde, 40*, 105–111.

Stanton, L. (2002). *Free some day: The African-American families of Monticello*. Thomas Jefferson Foundation.

Steinbrink, M. (2012). We did the Slum!: Urban poverty tourism in historical perspective. *Tourism Geographies, 14*(2), 213–234.

Steinecke, A. (2007). *Kulturtourismus: Marktstrukturen, Fallstudien, Perspektiven*. Oldenbourg.

Stone, P. (2005). *Dark tourism forum. University of Central Lancashire*. Retrieved from, www.darktourism.org.uk

Stone, P. (2006). A dark tourism spectrum: Towards a typology of death and macabre related tourist sites, attractions and exhibitions. *Tourism, 52*, 145–160.

Stone, P. (2011). Dark tourism: Towards a new post-disciplinary research agenda. *International Journal of Tourism Anthropology, 1*(3/4), 318–332.

Stone, P. (2012). Dark tourism and significant other death: Towards a model of mortality mediation. *Annals of Tourism Research, 39*(3), 1565–1587.

Stone, P., & Sharpley, R. (2008). Consuming dark tourism: A thanatological perspective. *Annals of Tourism Research, 36*(2), 574–595.

Stone, P., & Sharpley, R. (2009). *The darker side of travel: The theory and practice of dark tourism*. Routledge.

Strauss, M. (2001). Framework for assessing military dress authenticity in Civil War reenacting. *Clothing and Textiles Research Journal, 19*(4), 145–157.

Timothy, D. J. (2011). *Cultural heritage and tourism: An introduction*. Channel View.

Timothy, D. J., & Boyd, S. W. (2003). *Heritage tourism*. Prentice Hall.

Tomljenovic, R., & Faulkner, B. (2000). Tourism and world peace: A conundrum for the 21st century. In B. Faulkner, G. Moscardo, & E. Laws (Eds.), *Tourism in the 21st century* (pp. 18–33). Continuum.

Tunbridge, J. E. (2005). Commodifying the heritage of atrocity? Penal colonies and tourism. In G. J. Ashworth & R. Hartmann (Eds.), *Horror and human tragedy revisited: The management of sites of atrocity for tourism* (pp. 19–40). Cognizant.

Tunbridge, J. E., & Ashworth, G. J. (1996). *Dissonant heritage: The management of the past as a resource in conflict*. Wiley.

Urry, J. (1990). *The tourist gaze: Leisure and travel in contemporary societies*. Sage.

Vanneste, D. (2012, February). *Visiting battlefields: Darkness of the site or darkness of the mind? Insights from a visitors'/tourists' survey in the Flanders WW I Battlefield area*. Paper presented at the Annual Meeting of the Association of American Geographers, New York City, NY.

Vanneste, D., & Foote, K. (2009). *In Flanders field 2014: War, heritage, preservation, tourism, regional identity and the remembrance of WWI in Belgium*. Paper presented at the annual meeting of the Association of American Geographers, Las Vegas, NV.

Williams, S. (2009). *Tourism geography: A new synthesis*. Routledge.

Wolf, A., & Matzner, C. (2012). Arten und motive des dark tourism. In H. Quack & A. Steinecke (Eds.), *Dark tourism: Faszination des Schreckens* (pp. 89–100). University of Paderborn Press.

Young, K. (2009). Auschwitz-Birkenau: The challenges of heritage management following the Cold War. In W. Logan & K. Reeves (Eds.), *Places of pain and shame: Dealing with "difficult heritage"* (pp. 50–67). Routledge.

1 Extension to research note "dark tourism, thanatourism, and dissonance in heritage tourism management

New directions in contemporary tourism research" (*Journal of Heritage Tourism*, 2014, Vol. 9, No 2, 166–182) for the time period 2013–2023

Rudi Hartmann

The 2014 research note article (internet version June 2013) turned out to be one of the most widely read as well as most cited articles published in the *Journal of Heritage Tourism* (https://www.tandfonline.com/action/showMostReadArticles? JournalCode=rjht20 and https://www.tandfonline.com/action/showMostCited Articles?journalCode=rjht20). While the article was considered an interesting, quotable summary of conceptual developments in the heritage tourism field, dozens of articles and books presented in the literature *in the following years* are not covered. With this addition and extension, the attempt is made to include a review of publications that built on the conceptual foundations of what has been largely termed the "dark tourism" direction in a widening field of scholarly interest. Attention is given to selected studies published in the time period 2013–2023.

Probably, the most thorough review of dark tourism and thanatourism research in the heritage tourism field is Duncan Light's progress report, with a detailed analysis of the 1996–2016 literature (Light 2017). For a 20-year period, a quantitative assessment of work in this research area clearly showed that the number of studies has consistently increased from 1 to 5 papers per year during 1996–2001 to 20 papers in each 2015 and 2016. Light took a look at the aims and scopes of the published papers. He found that multiple efforts were made to clarify the parallel dark tourism and thanatourism conceptions as well as toward broadening the two terms' scopes. Two other distinguished foci of the studies were "understanding dark tourists" including their motives and the visitors' experiences and behavior. A large number of papers also discussed the management and marketing of places of death and suffering.

The Palgrave Handbook of Dark Tourism Studies (Stone et al 2018) is considered by editor-in-chief Philip R. Sone, a "landmark publication (in the wider dark tourism arena) and main literature resource for many years to come". The 768-page volume has 30 contributions ordered along six thematic sections: Section One – Dark Tourism History (edited by Tony Seaton), Section Two – Dark Tourism: Philosophy

DOI: 10.4324/9780367823795-3

and Theory (edited by Philip R. Stone), Section Three – Dark Tourism in Society and Culture (edited by Richard Sharpley), Section Four – Dark Tourism and Heritage Landscapes (edited by Rudi Hartmann), Section Five – The Dark Tourist Experience (edited by Philip R. Stone) and Section Six – The Business of Dark Tourism (edited by Leanne White). Two preeminent conceptual contributions in the Handbook are Tony Seaton's "Encountering Engineered and Orchestrated Remembrance: A Situational Model of Dark Tourism and Its History" (Seaton 2018) and Phil Stone's "Dark Tourism in an Age of 'Spectacular Death'" (Stone 2018). Among the many noteworthy articles are Erik Cohen's "Thanatourism: A Comparative Approach" (Cohen 2018), Mona Friedrich's "Dark Tourism, Difficult Heritage, and Memorialization: A Case of the Rwandan Genocide" (Friedrich et al 2018), Stephen Hanna's "From Celebratory Landscapes to Dark Tourism Sites? Exploring the Design of Southern Plantation Museums" (Hanna et al 2018) and Geoffrey Bird's "Marketing Dark Heritage" (Bird et al 2018).

Among edited volumes in the dark tourism field Glenn Hooper and John J. Lennon's *Dark Tourism: Practice and Interpretation* stands out (Hooper and Lennon 2017). Two articles of the book deserve to be highlighted: Ashworth and Tunbridge's "Death Camp Tourism: Interpretation and Management" (2017) and their provocative question in the chapter "Is all tourism dark?" (Tunbridge and Ashworth 2017). John Lennon, as a co-founder of the 'dark tourism' research direction, contributed new insights into the visitation and marketing of the Dachau Memorial Site (Lennon and Weber 2017). A further discussion of the visitation practices at the Rwanda genocide site is presented by Sharpley and Friedrich (2017). Finally, an emerging dark tourism research site, Rio de Janeiro's favelas, is covered by Hooper (2017).

Philip Stone's influence in the dark tourism research field can hardly be underestimated. He continues to have merits in dark tourism organizational initiatives as well as pushing new research questions to the forefront. Two largely neglected groups are given due attention: the female dead in dark tourism and the young tourists' experiences. Stone and Morton argue that, in death, men and women are not treated the same. The authors offer a critical review of how the female dead are portrayed in dark tourism with close-ups of Jack the Ripper's female victims and the sexualized corpses of the women of body worlds. They maintain that it is a male gaze of the female dead that dominates (Stone and Morton 2022). As to children and young people, they are seen but not heard in (dark) tourism research agendas. Stone examines the young tourists' experiences at dark tourism sites and proposes a conceptual framework (Kerr et al 2021). Two regional dark tourism explorations are presented as a book chapter ("Dark Tourism and 'Painful Pasts' in Africa: Concepts, Contexts and Challenges", Stone 2023) in *Cultural Heritage and Tourism in Africa* (Timothy 2023) and as a local monograph *111 Dark Places in England That You Shouldn't Miss* (Stone 2021).

Philip Stone also initiated a special section on the "History of Dark Tourism" in the *Journal of Travel History* (Hartmann et al 2018). The four discussants representing different scholarly perspectives in the dark tourism field took up Stone's point of departure: How have a variety of forces fuelled dark tourism in the past? Lennon argued that dark tourism has exhibited elements of the modern and post-modern. Further, he maintained that "visibility and media coverage, which has grown exponentially, has had some of the most significant recent impacts on consumer awareness and travel behavior following terrorist atrocities". Rice examined developments in Lancaster's slave trade history and its early commemoration in 1796. He warns that "in commemorating sites of African Atlantic presence it behoves us to be very careful not to be self-indulgent and not to be complicit in acts of appropriation that resemble the colonization of sentimentalization of the past". Noting Ricoeur's vision of the continued search for justice, he reminds us that

> a slave tour should not wallow in pain and trauma, but rather be a pedagogic praxis that enables action for change in the contemporary moment. If it does not at least aspire to this then what is its point but a pleasure principle for the entitled and privileged who get a vicarious thrill from the horrors of the past.

By looking at our neglect to consider genocides of the distant past, as committed for instance during the Egyptian and Roman empire periods, Reynolds asks the question: "Does the death depicted as a tourist site have to be recent to qualify as 'dark'?"

> My point is not to deny the unique features of death-focused tourism in the present. There is something qualitatively different about tourism in an age when creating and customizing travel experiences has never been easier, and when information, including graphic images and videos of disaster, can spread virally across the Internet.

Hartmann draws attention to the context in which the notion of dark tourism and the eventual formation of a new research tradition developed, that is in the spatial context of research in Southern Scotland and Northern England and within the intellectual framework of UK tourism studies. Prominent researchers helped to spread the new dark tourism perspective within the UK first – before the new concepts and perspectives in dark tourism "went global". Stone summarized in his response that Lennon, Rice and Reynolds have suggested "tourist sites of tragic history are places where public and vernacular (hi)stories and memories intersect and act in dialogue". While the focus of the round table discussion was on the history of dark tourism, it was equally an assessment of what dark tourism is in essence and what it could be in the future.

References

Ashworth, G. and Tunbridge, J. (2017). Death Camp Tourism: Interpretation and Management. In Hooper, G. and Lennon, J. (Eds.), *Dark Tourism: Practice and Interpretation*, pp. 69–82.

Bird, G., Westcott, M. and Thiesen, N. (2018). Marketing Dark Heritage: Building Brands, Myth-Making and Social Marketing. In *The Palgrave Handbook of Dark Tourism Studies*, pp. 645–663.

Cohen, E. (2018). Thanatourism: A Comparative Approach. In *The Palgrave Handbook of Dark Tourism Studies*, pp. 157–171.

Friedrich, M., Stone, P. and Rukesha, P. (2018). Dark Tourism, Difficult Heritage, and Memorialisation: A Case of the Rwandan Genocide. In *The Palgrave Handbook of Dark Tourism Studies*, pp. 261–289.

Hanna, S., Alderman, D. and Bright, C. (2018). From Celebratory Landscapes to Dark Tourism Sites? Exploring the Design of Southern Plantation Museums. In *The Palgrave Handbook of Dark Tourism Studies*, pp. 399–421.

Hartmann, R., Lennon, J., Reynolds, D., Rice, A., Rosenbaum, A. and Stone, P. (2018). The History of Dark Tourism. *Journal of Tourism History*, 10 (3), pp. 269–295.

Hooper, G. (2017). Dark Tourism in the Brightest of Cities: Rio de Janeiro and the Favela Tour. In Hooper, G. and Lennon, J. (Eds.), *Dark Tourism: Practice and Interpretation*, pp. 187–204.

Hooper, G. and Lennon, J. (2017). *Dark Tourism: Practice and Interpretation*. London: Routledge.

Kerr, M., Stone, P. and Price, R. (2021). Young Tourists' Experiences at Dark Tourism Sites: Towards a Conceptual Framework. *Tourist Studies*, 21 (2), pp. 198–218.

Lennon, J. and Weber, D. (2017). The Long Shadow – Marketing Dachau. In Hooper, G. and Lennon, J. (Eds.), *Dark Tourism: Practice and Interpretation*, pp. 26–39.

Light, D. (2017). Progress in Dark Tourism and Thanatourism Research: An Uneasy Relationship With Heritage Tourism. *Tourism Management*, 61, pp. 275–301.

Seaton, T. (2018). Encountering Engineered and Orchestrated Remembrance: A Situational Model of Dark Tourism and Its History. In *The Palgrave Handbook of Dark Tourism Studies*, pp. 9–31.

Sharpley, R. and Friedrich, M. (2017). Genocide Tourism in Rwanda: Contesting the Concept of the 'Dark Tourist'. In Hooper, G. and Lennon, J. (Eds.), *Dark Tourism: Practice and Interpretation*, pp. 134–146.

Stone, P. (2018). Dark Tourism in an Age of 'Spectacular Death'. In *The Palgrave Handbook of Dark Tourism Studies*, pp. 189–210.

Stone, P. (2021). *111 Dark Places in England You Shouldn't Miss*. Cologne: Emons.

Stone, P. (2023). Dark Tourism and 'Painful Pasts' in Africa: Concepts, Contexts and Challenges. In Timothy, D. (Ed.), *Cultural Heritage and Tourism in Africa*.

Stone, P., Hartmann, R., Seaton, T., Sharpley, R. and White, L. (Eds.). (2018). *The Palgrave Handbook of Dark Tourism Studies*. London: Palgrave Macmillan.

Stone, P. and Morton, C. (2022). Portrayal of the Female Dead in Dark Tourism. *Annals of Tourism Research*, 97, https://doi.org/10.1016/j.annals.2022.103506

Timothy, D. (2023). *Cultural Heritage and Tourism in Africa*. London: Routledge.

Tunbridge, J. and Ashworth, G. (2017). Is All Tourism Dark? In Hooper, G. and Lennon, J. (Eds.), *Dark Tourism: Practice and Interpretation*, pp. 12–25.

Part I: Introduction

Memorials of the Holocaust

The evolution of a new memorial landscape for the victims of Nazi Germany: the long and complicated path to the recognition of the former Nazi concentration camps as memorials and museums with interpretive centers

Rudi Hartmann

Few historical periods in human history are so fatally associated with the destruction of human lives as Hitler's 'Third Reich'. Historic places honoring the victims of National Socialistic Germany form a wide and expanding network of heritage sites in Europe. Most of the places where the horrific events occurred during 1933–1945 have been broadly denoted as Holocaust memorial sites in the remembrance of the six million Jews who died, and the many other ethnic, religious, social, and political groups which were subjected to persecution. Part I of the book reconstructs the evolution of this memorial landscape (Hartmann 2014, 2018).

A changing memorial landscape for the victims of National Socialistic Germany

During Nazi Germany's occupation of Central and Eastern Europe in 1941–1945, 20 main concentration camps, several extermination or death camps, and more than a thousand subsidiary or satellite camps were in existence (see, e.g., Gilbert's Atlas of the Holocaust 1982 and United States Memorial Museum's Historical Atlas of the Holocaust 1996). All the main camps, death camps, as well as hundreds of satellite camps have become memorial sites for the victims of National Socialistic Germany over the past decades.

The first memorial site was established at Majdanek near Lublin, Poland, in 1945/1946. It was here that the Allied Forces (Red Army) reached the first concentration camp in July 1944. As the Soviet forces moved very quickly in the direction of Lublin, the SS had little time to destroy or conceal facilities used in the mass murder of the prisoners – as they did, for instance, in the case of the early death camps of Belzec and Sobibor, which were inoperative by 1943. Thus, the physical

DOI: 10.4324/9780367823795-4

infrastructure of the Majdanek concentration camp found at liberation was largely unchanged and still had the gas chambers and the crematorium in place, as well as the storage of collected clothes and shoes of victims. Majdanek was the proof of what had been suspected about the nature of Nazi concentration camps in the early 1940s, and (Soviet) journalists visiting the camp shortly after made it public news (see, e.g. Gilbert 1982 and United States Holocaust Memorial Museum 1996). In November 1944, the Majdanek State Museum was founded by the Polish Committee of Liberation. It declared the camp a 'memorial site of the martyrdom of the peoples of Poland and other nations' (Marcuse 2010, pp. 192–193) which became accessible to the public in 1945–1946. It is estimated that 300,000–400,000 people visited the museum and site during the first two years (Jalocha and Boyd 2014). By 1947, the Polish Parliament passed a decree that the remains of the Majdanek camp site (jointly with those at Auschwitz and other concentration camps on Polish territory) were to be preserved. In 1965, Majdanek received the status of a national museum. However, Majdanek, as the second-largest concentration camp in Poland, would remain in the shadow of Auschwitz which became a leading symbol of the Holocaust (Hartmann 2018).

Historian Harold Marcuse reconstructed in great detail what happened to all the former main concentration camps, the prisoners, and the SS guards in the immediate years after liberation (2010). He lists five uses of the camps. First, the Allied Forces who were confronted with horrific atrocities when reaching and liberating the camps took measures to educate the populations living in the towns nearby such as Bergen-Belsen or Dachau about the conditions they found. Second, there was an urgent need to bring tens of thousands of survivors back to health. A third use was directed to imprison the Germans who were held responsible for the crimes committed at the sites. Thus, former camps like Dachau became the place where SS guards and others were kept in captivity while the trials proceeded. Fourth, efforts were made to preserve components of the camp environment which were considered important for future educational purposes. Finally, Marcuse reviews the lack of attention given to the more remote camps in the concentration camp system as well as death camps such as Belzec and Sobibor in Eastern Poland. These sites as well as the majority of the satellite camps were simply abandoned and ignored before they were included in the commemoration practices much later, in the 1960s and 1970s, and some as late as in the 1980s and 1990s (Marcuse 2010).

In the absence of accessible and inoperative memorial sites, it was the camp liberation anniversaries in the 1950s and 1960s that had importance for the former prisoners who vividly remembered liberation which marked a turning point in their lives. The dates of liberation for the larger camps – Buchenwald on April 11 (1945), Bergen-Belsen on April 15 (1945), and Dachau on April 29 (19435) – became major annual events which brought thousands of former prisoners together. Moreover, the gatherings at the early camp liberation anniversaries served as a forum for the discussion of how to establish first memorials, markers, and exhibits on the grounds. While the number of surviving concentration camp prisoners has dwindled over the past decades due to natural attrition, anniversary events are still held. In 2005, the United Nations General Assembly resolution 60/7 recognized the liberation of

Auschwitz on January 27 (1945) as *International Holocaust Remembrance Day*. It commemorates the genocide that resulted in the death of an estimated six million Jewish people, two million Romani people ('gypsies'), 250,000 mentally and physically disabled people, and 9,000 gay men by the Nazi regime and its collaborators (Hartmann 2018).

The first permanent memorial at a concentration camp in Germany was established in Bergen-Belsen. A collective Jewish monument was established in September 1945. On the first anniversary of the camp liberation on April 15, 1946, a stone monument with Hebrew and English inscription was inaugurated by the Central Jewish Committee of the British Zone. In 1947, efforts were started to create a central memorial in the form of an obelisk and memorial wall naming 14 nations of the victims in Belsen. The memorial was formally dedicated in a commemorative ceremony in 1952 attended by West German president Theodor Heuss and the President of the Jewish World Congress Nahum Goldman. The Bergen-Belsen camp was liberated by British and Canadian troops who found horrific conditions at the site. It is estimated that more than 70,000 people died at the POW and the Concentration Camp before and during the immediate weeks following liberation. A typhus epidemic raged during the final stage of the camp, and thousands of corpses of diseased prisoners were buried in nearby mass graves. Shortly after liberation, the camp grounds had to be completely cleared for health reasons. The uncontestable, widely reported magnitude of the fatalities in Belsen, the presence of a large nearby community of displaced persons, many of them survivors of the camps, as well as the complete removal of the structures on the grounds facilitated the allocation of the memorial. This may have contributed to a relatively fast decision for a memorial and the later approval by the State of Lower Saxony in charge of the site by 1952. The memory of the young author Anne Frank, who died with her sister, Margot, in Belsen in March 1945, gave further momentum to the memorial site in the mid-/late 1950s. The Bergen-Belsen memorial site saw more changes in the 1960s and the following years, from the addition of a small 'document house' in 1966 to the development of a new memorial site museum which opened in 2007 (Marcuse 2010).

In the case of Dachau, where at least 40,000 people died during the 12 years the camp existed, the push for a memorial site played out at a much slower pace and in more complicated ways. Early initiatives for a memorial turned out to be failures. Several proposals were turned down for a variety of reasons or forgotten by the public (Marcuse 2010). Local initiatives and plans for the closure and the demolition of the crematorium (with a first exhibit about the camp) in the early 1950s were prevented by the Paris Treaty, which West Germany had signed with France in 1954. Several clauses in the treaty protected the burial sites of the concentration camp prisoners and access to the camp. After 1955, it was most of all the re-founded Comite International de Dachau prisoner organization, which tenaciously stood up for the preservation of the camp site. The official Dachau Concentration Camp Memorial, with a museum and a small salaried staff funded by the State of Bavaria, was opened to the public in 1965. Despite the formal establishment of the Dachau memorial site, considerable resistance among the Dachau

residents persisted (Hartmann 1989; Marcuse 1990, 2001, 2005, 2010). People living in Dachau and the County of Dachau had a hard time coming to grips with the
fact that the first concentration camp of Nazi Germany was established next door
to their market town. It was a new generation of Dachau citizens and elected politicians who sought a more constructive relationship in the 2000s/2010s. In addition, many of the museum exhibits were redesigned, with other memorials created
in town. A new visitor center, with joint responsibility by the Memorial administration and the City, was established, and more programs in collaboration with
local Dachau historian docents were developed (Burger 2017; Hartmann 2017).
The Dachau Memorial site has become a major tourist destination with annually
900,000 visitors (Hammermann 2021), only surpassed by annual visitation at the
Auschwitz Memorial and the Anne Frank House in Amsterdam with 1.3 million
persons each.

The agenda and management of the memorial sites for the victims of Nazi Germany have seen major changes as can be reconstructed for the memorial sites in
former East Germany including its paramount site, the Buchenwald National Site
of Commemoration and Warning ('Nationale Mahn- und Gedenkstaette'), where
more than 50,000 people died during 1937–1945. The site was opened in 1958,
with a new memorial – away from the 'beach tree' forests ('Buchenwald') where
the concentration camp was hidden – now facing toward the valley floor and the
City of Weimar. The expansive new memorial consisted of a wide sloping walkway,
along seven bas-reliefs which showed the plight and ultimately successful struggle of the prisoners against Fascism, to a large gathering place lined by an avenue
of 18 featured nations with a series of massive pylons. The new memorial was an
impressive backdrop for political action. It was at the Buchenwald memorial site
where soldiers and young party members took their oath, that school classes from
all over the DDR came to learn about the victory of the German Communist movement and the continued successes of the East German State. How did the political
changes after 1989/1990, with the fall of the Berlin Wall and the re-unification of
Germany, affect the Buchenwald site and its management? Free elections in the
former East German states brought significant changes to the administrative body
of the memorial sites. A democratization in the decision-making processes, most of
all with the inclusion of victim groups so far neglected, resulted in new guidelines
and policies. The outcome was the decision for a fundamental reorientation of the
memorial site to the commemoration of the victims (rather than a celebration of
the successful fight against Fascism) and from the monumental memorial site back
to the former concentration camp. Several memorials were added to the grounds
of the former camp, among them the Jewish Memorial (1993) and a Memorial for
the Sinti and Roma (1997). In 2002, a memorial for the victims at the Little Camp
('Kleine Lager'), where several thousand Jewish lives perished during 1944–1945
under horrific conditions is described by Buchenwald survivor Elie Wiesel. The
main museum exhibits were redesigned and a new site was added that focused on
the history of the memorial site itself. Most controversial was the establishment of
a memorial and museum at the Special Camp Nr. 2, which was in existence during
the Soviet occupation of the camp 1945–1950. Seven thousand persons died there,

mostly members of the SS and the NSDAP Nazi party who had functions at the concentration camp as well as socialists who fell out of favor in the early years of Communism in East Germany. While 'winds of change' blew across the Buchenwald memorial site in the 1990s, resulting in a different political culture, all the memorials, monuments, and markers were kept in place (Stiftung Gedenkstaetten Buchenwald und Mittelbau-Dora 2003).

Permanent changes in the management of the sites came with a new administrative organization for the larger memorial sites in both the new states (in former East Germany) and the West German states during the 1990s and 2000s. New foundations for the administrative support of the memorial sites were formed on a state level (within ministries in charge of cultural affairs). These state-supported agencies gave the memorials financial and personal stability at last. A general consensus emerged in re-unified Germany that supporting the memorial sites was a crucial public responsibility to be sufficiently and consistently taken on (Hartmann 2018).

New sites, new forms of commemoration, new perspectives

There is no city in Germany with so many new memorial sites in honor to the victims of National Socialism as in Berlin. Among the prominent new sites are the memorial 'Topography of Terror' at the former Gestapo Headquarters (1987, with new documentation center 2010), the Memorial to the Persecuted Homosexuals under National Socialism (2008), the Memorial to the Sinti & Roma Victims of National Socialism (2012), and the Memorial for the 'Euthanasia' Murder Victims (2014). There are also memorials to uprisings and resistance, in West Berlin the memorial and museum at the Bendlerblock and at the Ploetzensee Memorial, in East Berlin near the Dom a memorial for the resistance group of Herbert Baum, and near Alexander Platz the memorial for the 'women of the Rosenstrasse' demonstrating for the release of their Jewish husbands. A uniquely designed memorial was established at the site of the Nazi-orchestrated book burning on May 10, 1933, with a view into empty bookshelves. Along the Unter den Linden boulevard in the Berlin Mitte district is the New Guardhouse (Neue Wache), the Central Memorial of the Federal Republic of Germany of the victims of war and dictatorship, with the Kaethe Kollwitz *Pieta Mother with Dead Son*. Finally, the most remarkable (and hotly debated) memorial was inaugurated in 2005, 60 years after the end of WWII, near the Brandenburg Gate in the center of the City: the Memorial to the Murdered Jews of Europe. It shows 2,177 steles of different heights on an open square; placed underneath is the formal museum 'Ort der Information' with the names of three million victims of the Holocaust. The widely visited, freely accessible monument – possibly the most visited Holocaust memorial in Europe – has remained a structure with ambiguous meanings for many (Freudenheim 2005; Petrow 2005).

A new form of commemoration was proposed by the Cologne artist Gunter Demnig in 1992, who came up with the idea of placing memorial markers in front of the last residence of victims of the Holocaust. The 'Stolpersteine/Stumbling Blocks' had the form of a ten-centimeter concrete cube bearing a brass plate with an inscription of the victim's name and life dates, including when they were born and deported and where they were murdered and died. Gunter Demnig's innovative agenda caught on and led to a multiplicity of community events in hundreds of towns and cities. As of June 2023, there were 100,000 stumbling blocks in 22 countries of Europe, including in big cities like Berlin (with more than 10,000 markers) and small towns like Dachau (with ten stumbling blocks).

Among the new perspectives which can be distinguished for reminding the public of the events of 1933–1945 in a broad sense are the memorials for *victims and persecutors* ('Opfer' und 'Taeter'). Most often, the new memorials at 'Persecutor Sites/Taeter-Orte' are equipped with documentation exhibits and education centers. Prominent (and highly visited) places in this context include Hitler's mountainside retreat near Berchtesgades in the Bavarian Alps, with a new Obersalzberg Documentation Center since September 2023, and the Nazi Party rally grounds in Nuremberg. Here, the rally grounds used for the annual mass events from 1933 to 1938 have seen multiple post-WWII uses as well as the development of exhibits. Another persecutor-centered place, in Berlin, is the Wannsee Villa (Haus der Wannsee-Konferenz), where the SS decided on the 'Jewish final solution' on January 20, 1942. It has become a place of memory and education (1992) (see the special issue 'Gedenkstaetten an NS-Taeterorten', Geschichte in Wissenschaft und Unterricht 2021).

Thousands of memorial sites, as major places of commemoration with visitor centers and museums or as small sites with simple plaques and markers, have been established in many European countries. They include Auschwitz II Birkenau, where more than a million people died, and the Babyn Yar Massacre Site near Kyiv, Ukraine, with 33,771 men, women, and children murdered, as well as creatively designed memorials at the Treblinka extermination camp (see appraisal 'as perhaps most magnificent of Holocaust memorials' by Young 1993) and at the 'Shoes on the Danube Bank' Memorial in Budapest, Hungary.

Still, many memorial sites are in the development stage or have remained incomplete as the aforementioned Holocaust memorial at Babyn Yar (Hammer 2023). One of the documentation centers in the planning phase is the memorial at the Kaufering satellite camp complex (in existence for ten months in 1944/1945) in or near Landsberg. A scholarly concept for the exhibits has been completed and waits to be implemented in the near future (see Chapter 3). Another area of slow recognition is the numerous sites connected to the German aviation armament industry which was an important agenda in the late phase of WWII and saw the horrendous, fatal uses of mostly Jewish prisoners (see Chapter 4). In the Netherlands, existing sites are restructured and remodeled to serve the continued interest of the public during the long period of Nazi occupation (1940–1945). New research and approaches in museum technology will see its application for the changed memorial sites (see Chapter 5). Even in Dachau, new additions to the memorial and updates to existing facilities and exhibits are in the planning or implementation phase (Hammermann 2021; see Chapter 2). All this is proof of a still-expanding Holocaust memorial landscape in Europe, with shifts and re-interpretation as to its contents and presentation.

References

Burger, W. (2017), *Learning – remembering – meeting: Educational work at the Dachau concentration camp memorial site*, Dachau: KZ-Gedenkstaette Dachau/Stiftung Bayerische Gedenkstaetten.

Freudenheim, T. (2005), Monument of ambiguity, *The Wall Street Journal*, June 15.

Geschichte in Wissenschaft und Unterricht. (2021), *Schwerpunkt: Gedenkstaetten an NS-Taeterorten*, Vol. 72, Issue 3–4, 2021, 120–248.

Gilbert, M. (1982), *Atlas of the Holocaust*, New York: Macmillan.

Hammer, J. (2023), The ravine: Plans to create a uniquely ambitious Holocaust memorial at the site of an infamous massacre were finally underway. The a new war in Ukraine changed everything, *Smithonian*, Vol. 54, Issue 6, December, 66–79.

Hammermann, G. (2021), Die KZ-Gedenkstaette Dachau – Zukunft der Erinnerung, *Geschichte in Wisssenschaft und Unterricht*, Vol. 72, Issue 3–4, 125–144.

Hartmann, R. (1989), Dachau revisited: Tourism to the memorial site and museum of the former concentration camp, *Tourism Recreation Research*, 41–47. Reprinted in *Tourism Environment*, Tej Vir Singh, et al. (Eds.), New Delhi Inter Indian Publications, 1992, 183–190.

Hartmann, R. (2014), Distinguished visitor keynote address "Tourism to memorial sites of the Holocaust: Changing memorial landscapes, changing approaches to the study of the sites associated with the victims and perpetrators in Nazi Germany" at Symposium "Assessing dark tourism scholarship: New frontiers" organized by the Institute for Dark Tourism Research, University of Central Lancashire, Preston, UK, September 23.

Hartmann, R. (2017), Places with a disconcerting past: Issues and trends in Holocaust tourism, *EuropeNow*, Issue 10, September 6, 6–12.

Hartmann, R. (2018), Tourism to memorial sites of the Holocaust, in *Palgrave Handbook for Dark Tourism Studies*, Rudi Hartmann (Co-Ed.), Jointly with P. Stone, T. Seaton, R. Sharpley and L. White, London: Palgrave Macmillan, 469–507.

Jalocha, M. and Boyd, S. (2014, August 13–17), *Tourism development opportunities for the Lublin region of Poland: Emphasis beyond dark heritage*, Paper presented at tourism and transition in a time of change, Pre-Congress Meeting Krakow/Pieniny Mts, Poland.

Marcuse, H. (1990), Das ehemalige Konzentrationslager Dachau: Der muehevolle Weg zur Gedenkstaette, 1945–1968, *Dachauer Hefte*, Vol. 6, 182–205.

Marcuse, H. (2001), *Legacies of Dachau – the uses and abuses of a concentration camp, 1933–2001*, New York: Cambridge University Press.

Marcuse, H. (2005), Reshaping Dachau for visitors: 1933–2000, in *Horror and human tragedy revisited – the management of sites of atrocities for tourism*, G. Ashwortrh and R. Hartmann (Eds.), New York: Cognizant Communication Corporation, 118–148.

Marcuse, H. (2010), The afterlife of the camps, in *Concentration camps in Nazi Germany*, J. Kaplan and N. Wachsmann (Eds.), New York: Routledge, 186–211.

Petrow, C. (2005), Memorial to the murdered Jews of Europe, Berlin. *Topos*, Vol. 50, 86–92.

Stiftung Gedenkstaetten Buchenwald und Mittelbau-Dora. (2003), *Ueberlebensmittel Zeugnis Kunstwerk Bildgedaechtnis – Die staendige Kunstausstellung der Gedenkstaette Buchenwald – Denkmale auf dem Lagergelaende*, Weimar: Stiftung Gedenkstaetten Buchenwald und Mittelbau-Dora.

United States Holocaust Memorial Museum. (1996), *Historical atlas of the Holocaust*, New York: Macmillan.

Young, J. (1993), *The texture of memory – Holocaust memorials and meaning*, New York: Yale University Press.

2 The memorial site at the former Dachau concentration camp (1933–1945)

A dissonant heritage for a small Bavarian market town and city which has become an internationally recognized destination[1]

Rudi Hartmann

The concentration camp in Dachau was established in March 1933 shortly after the Nazi seizure of power. It was among several dozens of early camps but would be the only concentration camp in operation for the 12 years of Nazi rule in Germany. The SS in charge of the prison facilities kept more than 200,000 prisoners there; initially, the camp inmates were opponents of the 'Third Reich', largely Communists, Social Democrats, Labor Unionists and members of the confessing churches. Eventually, they comprised all groups not fitting according to Nazi ideology into the German People's Community ('Volksgemeinschaft'), including Jews (they made up about 25% of the prisoners), Sinti and Roma ('gypsies'), Jehovah's Witnesses, homosexual men and 'asocials'. Besides 32,000 inmates from the German Reich, thousands of prisoners of other nationalities arrived in Dachau: 40,700 from Poland, 25,300 from the Soviet Union, 21,300 from Hungary, 14,100 from France and 9,600 from Italy. Altogether more than 40 nationalities were present at the camp grounds. Hundreds of prisoners, many of them children, were subject to medical experiments and died. More than 2,000 incapacitated prisoners who could no longer work fell victim to the Euthanasia programs. In 1945, thousands of prisoners in the overcrowded camp were affected by a typhus epidemic. It has been reconstructed that 41,500 people died in or near the camp during the years 1933–1945 (Hammermann 2021).

Dachau was considered a 'model camp'. SS Colonel Theodor Eicke, camp commandant in June 1933, was promoted to leader of the whole concentration camp system, consisting, by the end of the war, of 20 to 25 main camps and over 1,200 subsidiary camps. Eicke developed a tightly organized administrative and command structure with a set of disciplinary and punishment regulations to be enforced on the prisoners. Many of the camp commandants, such as Rudolf Hoess in Auschwitz, started their careers in Dachau.

On April 29, 1945, the Dachau concentration camp, then holding 32,000 prisoners in the main camp, was liberated by American Forces. In the first few weeks, it was used for sick persons to be nurtured back to their recovery. Then

DOI: 10.4324/9780367823795-5

it served as the internment camp for war criminals and perpetrators from the SS and other Nazi organizations which were held there for the 'Dachau Trials' 1945–1948 (Hammermann 2003). Eventually, the camp would become a housing area for displaced persons and ethnic Germans from Eastern Europe who had lost their homes (Marcuse 2010). Only a small confined area near the crematorium in the camp was used for the commemoration of the tragic events. In 1950, the sculpture of the Unknown Concentration Camp Inmate was placed there. A first exhibit about the Dachau KZ entitled *Never Again* – then changed to *Remember That* to avoid reproaches the documents would stir up bad feelings – was shown in the crematorium. Following criticism the exhibit was removed by state authorities on May 12, 1953 (Marcuse 2001, 170–185). Plans to close public access to the crematorium and to eventually tear it down were in discussion in 1953–1955.

The residents of the Town of Dachau which were confronted with the horrible deeds right 'in their backyard' in the early post-war period asserted to the United States. Agencies that they knew nothing about the crimes in the camp. They showed little interest in helping to create a memorial site – neither were the leading powers in government and parliament within the State of Bavaria in the 1950s. It took 20 years till the Dachau Memorial Site was established and opened on May 9, 1965 (Marcuse 1990). There were mainly two circumstances that led to the foundation of the memorial site. First, the 1955 addition to the Paris Treaty 'ensured unhindered access' to the grave sites and places where the lives of the victims of Nazi Germany had perished, including in Dachau (Marcuse 2001, 185). Thus, initial local/state considerations of demolishing the camp site or major parts like the crematorium had no longer a legal basis in view of the international treaty obligations. Second, there were the camp survivors who consistently pushed for a formal memorial site. At the tenth anniversary of the camp liberation in 1955 they re-founded the Comite International de Dachau (CID), which became the driving force for a Dachau Memorial Site. By 1959/1960, main officials of the Catholic Church, most prominently the Munich suffragan Bishop Johannes Neuhaeusler, openly supported a memorial site, and at the occasion of the Eucharistic World Congress, a first memorial was inaugurated on the camp grounds: the Mortal Agony of Christ Chapel in 1960 and in 1964 a Carmelite Convent. Several memorials followed: the Protestant Church of Reconciliation (1967), the Jewish Memorial (1967), and a Russian Orthodox Chapel (1994). The International Memorial designed by Holocaust survivor Nandor Glid was dedicated on September 9, 1968. It continues to be a central symbol for Dachau as a place of memory. A multitude of information plaques and several new structures, including two rebuilt barracks, have been added at the Dachau Concentration Camp Memorial Site. A number of memorial sites were established outside the camp, including at the Leitenberg Cemetery, the Waldfriedhof Cemetery, and a memorial at the Hebertshausen Shooting Range. See a detailed list of sites and structures with the respective specific historic information in a current *Tour of the Dachau Concentration Camp Memorial* (Hammermann and Pilzweger-Steiner 2017, 88 pages).

2.1 From dissonance to acceptance in Dachau

After the official opening of the Dachau Concentration Camp Memorial Site in 1965 conflicts and tensions between the City of Dachau and the Dachau Memorial Site were a defining part of the early relationships, until the mid-1990s. Resentments over the wide attention Dachau received internationally because of the concentration camp years were a dominant feeling among Dachau's residents. Mayor Hans Zauner (1933–1945 and 1952–1960) considered himself as one of the 'good Nazis' when he took office, whereas he maintained at the same time that those with questionable characters were among the concentration camp prisoners (Marcuse 2001, 79–81). Mayor Lorenz Reitmayer's (1966–1996) main agenda was to represent and promote 'the other Dachau'. At the 25th anniversary of the liberation of the Dachau Concentration Camp 1970, Reitmeier expressed his feelings at the Dachau Council Meeting as follows: "The establishment of the concentration camp has achieved in twelve years, to destroy everything that Dachau had acquired in good reputation in centuries" (quoted in Hartmann 1989). Both mayors, which were freely elected and popular in town during and after their tenure, fostered dissonance towards the leadership and administration at the concentration camp memorial site. Mayor Reitmeier never gave a talk at the memorial site during his 30-year tenure (Marcuse 2001, 329–333).

Animosities between right-wing Dachau politicians and left-leaning members of the CID and the memorial site had the consequence of a lack of information exchange between the city and the memorial site. For long, the city's information policy largely ignored the many events held at the memorial site. For extended time periods, the local tourist office did not show or hand out materials prepared by the memorial site (Hartmann 1989). Responsibilities and work conditions for Barbara Distel, a long-time leader at the memorial site (1975–2008), were affected by these tensions, while she faced various challenges in and outside Dachau (Marcuse 2001; Benz 2008; Benz, Distel and Koenigseder 2009). Though, information about the Dachau locale and the works of Dachau residents were not represented at the memorial site. For extended time periods, only the *Konzentrationslager Dachau* exhibition catalog (in several languages) was available for sale to the tourists (Distel and Jakusch 1978). The works of the Dachau journalist and book author Hans-Guenter Ricardi, for instance, were ignored. He published an important early study about the Dachau Concentration Camp 1933/1934 (Richardi 1983) and later presented with other Dachau locals a useful guide to Dachau's contemporary history with chapters about politics in Dachau 1915–1933 and most remarkedly a feature on resistance fighters in the late days of the camp in April 1945 (Richardi, Philipp and Luecking 1998). The insufficient literature situation at the memorial site would fundamentally change with the opening of the new visitor center 2009 equipped with a front desk with both information materials from the memorial site and the city and a book store with a wide assortment of literature.

How did the deplorable situation change? It was a slow process that started in the mid-/late 1980s and early 1990s. A group of local historians in town formed to do research on the city and the camp. A first exhibit was shown in the lobby of the City Hall (Zum Beispiel Dachau 1983); several publications and leaflets followed in the late 1980s and the 1990s. Further, an extended debate over the establishment of an International Youth Guesthouse in Dachau in 1998 contributed to the break-up of the fronts. The international hostel became eventually the Max Mannheimer House Study Center. Holocaust survivor Mannheimer's role as a time witness (lecturing at local schools and at the memorial site) and revered communicator between the memorial site and the city (from 1986 to 2016, the year he passed away at age 96) helped also considerably (Mannheimer 1985, 2008, 2009). For many years, he served as the chair of two main prisoner organizations, of the *Lagergemeinschaft Dachau* and as the vice-chair of the *CID*. On the government level, incoming mayor Kurt Piller (1996–2002) represented a contrast to previous mayor Reitmeier, as a person who had visited Israel and frequented the memorial site before taking office. Last but not least, the Bavarian prime minister Max Streibl paid his first official visit as the head of the State of Bavaria in 1993.

A major update and re-organization of the memorial site 1995–2003, of the main exhibition, and the rerouting of the entrance to the Concentration Camp Grounds, through the Gate at the Jourhaus (the way the prisoners entered the camp), contributed to a more adequate rendering of the camp's history and to a more inclusive image of all prisoners at the site. Forgotten prisoner groups like the gay men and the Sinti and Roma found at last due attention at the memorial site. A new visitor center which was opened in 2009 made also a big difference. Here, the memorial site and the city have had a collaborative role. Finally, it should also be mentioned that the Dachau Memorial Site was put under the umbrella responsibility of a newly established foundation, the *Stiftung Bayerischer Gedenkstaetten*, in 2003 (Bavarian Memorial Foundation 2023). It made the administrative process of commemorating the victims of Nazi Germany in the State of Bavaria more transparent.

Quite important for the improvement of the relations between the residents of Dachau and the memorial site was the establishment of a *Path of Remembrance* 2007, with a dozen marked stops, plaques, and photographic records, which showed the path prisoners had to take from the railway station to the concentration camp, and the installing of *Stolpersteine* in town in 2005/2014/2017. Six of the first Stolpersteine (brass-covered 'stumbling stones' that reminded residents of fellow Jew citizens murdered in concentration camps) were laid in front of their homes in 2005, then others for politically persecuted locals and the people who died in the '14 f 13' actions of 'Euthanasia'. Like the new International Youth Guest House, the Path of Remembrance and the Stolpersteine were set up in residential areas of Dachau.

A significant step in improving the memorial site's interaction with the visiting public was the establishment of an Educational Department in 2001. By 2010, it had four permanent staff members who developed programs like monthly lectures on a given theme at the site as well as within the context of the city (Burger 2017). The memorial site started to employ local lecturers ('Referenten') who – after educational coursework – were given a chance to talk about relevant topics and issues of contemporary history from a Dachau angle (Hartmann 2017). Twenty-five themed walks & talks led by local lecturers were organized for the year 2021 (Dachauer

Gaestefuehrer 2021). A large number of these local lecturers are now in a part-time employment situation with the memorial site (Schwenke 2022).

2.2 Tourist numbers

Visitation of the new memorial site, from 1965 to 1975, was considerable from the very beginning. 'Dachau' was among the most highly visited sites in the Munich Metro Area/Upper Bavaria (Hartmann 1989, 2004; Marcuse 2005). On an annual average, 360,000 people came to see the Dachau Concentration Camp Memorial site, a place hardly advertised in tourist ads or mentioned in guidebooks. Ruth Jakusch, a former CID employee, was in charge during these early formative years. It should be mentioned that entry to the memorial site was free of charge and that the given numbers were estimated figures. This applied also to the given numbers of visiting West Germans and the international visitors, with the latter leading visitation during the 1960s/1970s/1980s. It were largely American tourists[2] as well as US soldiers stationed in Germany who came, in addition to visitors from neighboring European countries and from Israel. From 1975 on, the number of tourists to the Dachau Memorial Site rose constantly, to about 900,000 in the mid-/late 1980s. Marcuse argues attitudes in West German society had changed, with a new crop of young teachers who had less problems with the Nazi past. In significant numbers, they brought their students to 'Dachau'. It was also the *Holocaust* mini-series in 1979 (shown with great success in Germany in 1980) that made a significant impact. Interest in the once-dreaded chapter of the Nazi years now spiked (Marcuse 2001, 2005).

During the mid-/late 1980s visitation to Dachau may have reached or surpassed one million. The methods of counting the visitors were more and more flawed as they did not consider the changing visitor hours and a tendency among the visitors to omit the visit of the over-run main museum where the estimates were taken. The Dachau Concentration Camp Memorial Site was severely understaffed in the 1980s,[3] with fewer than 20 employees, compared, for instance, to the (East German) Buchenwald Concentration Site, with about 500,000 visitors (many of them new Communist Party members to be sworn in) and over a hundred employees (Hartmann 1989; Marcuse 2001, 2005).

In the 1990s/2000s visitation to the memorial site dropped slightly to 700,000/600,000. It could have been the result of the new travel possibilities after the 'Iron Curtain' fell in 1990, with the concentration and death camps in Poland becoming more easily accessible. Moreover, Marcuse argues that "Auschwitz has eclipsed Dachau as the most widely recognized symbol of Nazi atrocities" (2005, 118). In terms of visitation, the Auschwitz Memorial Site started to surpass Dachau. In the 2010s, 1.3 million visitors annually went to 'Auschwitz', a fairly accurate figure as they had to pay an entrance fee at the memorial site.

Currently, the number of visitors at the Dachau Concentration Camp Memorial Site has risen steadily from 600,000 to 900,000 (Hammermann 2021, 130–132). Still, foreign visitors are leading over Germans as do individual tourists (55%) over school classes (45%). More reliable counting methods, at the parking where car owners and bus operators arrived and had to pay a fee, and other observations made at places like the visitor center provide sufficiently long-term trends in visitation.

2.3 Issues and problems

While administrative conditions at the memorial site have been improved in many ways since 2010, resulting in a consolidation of the status of this institution, some perceptional problems in the wider public persist. 'Dachau' and other memorial sites have often been seen in negative lights by many, including extreme party groups in Germany. In particular, in the new states of former East Germany anti-democratic traditions survived and/or were revived and made headways in other regions and states. In Bavaria, the right-wing extremist party AfD (Alternative fuer Deutschland) gained considerable support in state-wide elections, in 2018 with 22 seats in the Parliament (according to a 10% share of the votes) and in 2023 with 32 seats (14.6% share of the votes). Not surprisingly, the Dachau Memorial Site had to witness neo-Nazi incidents. Intellectually, a historical revision has become more acceptable, at least for active supporters of these extremist factions. If political trends continue, the AfD may ultimately get into the position to impact personal and budget decisions in the State of Bavaria.

A shocking event occurred in November 2014. Overnight, somebody took out the heavy gate at the Jourhaus with the infamous inscription 'Arbeit Macht Frei' (Work makes you free). Two years later, in December 2016, it was found in Bergen, Norway, and was returned. During the period it was missing, the memorial site decided to replace it with a copy. The administration left the copy in the gate and used the returned original for an artifact in the main exhibit.

Figure 2.1 Jourhaus with visitors.

Source: photo by Rudi Hartmann

Figure 2.2 The Gate with the inscription 'Arbeit Macht Frei'.

The Jourhaus, with the gate, represented a core area of the concentration camp. It was part of the newly expanded camp built in 1938. This is where the new prisoners arrived, where they had a first roll call and 'welcome'. The two-story building housed the offices of the SS in charge for the day, with guards equipped with machine guns in the tower at night. Sadly enough, it was also on the way to the last passage of a dead prisoner – to the crematorium (Zamecnik 2004). So far, the Jourhaus has only two markers, with the information that American troops liberated the camp on April 29, 1945. The crucial offices for disciplinary hearings are now empty. It was at the Jourhaus that perpetrators and victims interacted. Can this outstanding locale on the camp grounds be integrated for visitation one day in the future?

2.4 The future of commemoration at the memorial site

In 1943/1944 the Dachau Concentration Camp appeared as a huge military-industrial complex from aerial photographs. It covered large areas of the City of Dachau. Population in the eastern sections surpassed by far the population living in the older, established areas of the town. The current memorial site shows only a portion of the whole complex.

When the Minister Council of the Bavarian Government met to develop a 'comprehensive concept for the commemoration of the events' (Gesamtkonzept fuer die Erinnerungskultur) on January 21, 2020, it was decided to include some of these still neglected areas (see Hammermann 2021):

- Remnants of the early concentration camp (1933–1937) where the rigid instructions of the 'Dachau model' were implemented and practiced
- The SS Camp (which made up the largest section of the Dachau Concentration Camp, not accessible now to the public), with historic buildings like the

commandant's headquarters (Kommandantur), the former camp bakery, and the workshops (Werkstaetten) where prisoners were forced to work
- The SS experimental agricultural facility/'herb garden'/'plantation' northeast of the camp, one of the most feared job areas for prisoners
- The internment camp for the Dachau Trials 1945–1948 (where the members of the Dachau SS, of the SS working in the Flossenbuerg, Mauthausen, Buchenwald, Muehldorf, and Mittelbau-Dora concentration camps were held for 489 proceedings at the military court) northwest of the camp

As some of the areas are currently used by state agencies (for instance, the SS Areal by the Bavarian Police), other features of the comprehensive renewal concept of the memorial site are given preference. A new documentary film has been released (Hammermann, Thomas, Pilzweger-Steiner, Schweitzer, Meyer-Krahmer and von Wedemeyer 2020). It was the first step in the renewal of the museum exhibit.

The central main exhibit will be revised not only with an update of its contents but also regarding a rapidly changing museum technology, with new requirements. The perspective of an 'integrated history' which shows events from the angle of the victims, the perpetrators, and the bystanders is planned to be introduced. Other exhibits, at the Concentration Camp Prison (Bunker), at the crematorium, at the Leitenberg Cemetery, at the 'Herbal Garden'/'Plantation' areas, and at the workshops, will be also revised, expanded, and/or added on.

Some of the structures on the camp grounds, such as the two barracks and the Jourhaus, will be renovated for visitation.

One of the great challenges for a comprehensive renewal of the memorial site is the fact that time witnesses are no longer available and that a younger generation of visitors requires new methods of communication. A major project is the production of virtual tours of the memorial sites (Schwenke 2022). Some of the smaller tours, for example, a tour for visiting children or a tour for members of the German military (Bundeswehr), are already in place. Moreover, at specific sites (for instance, at the site where the Politische Abteilung/Department of Political Affairs was housed) short tours are available on the internet. Currently, ideas and conceptual plans for a wider narrative (or narratives) are under review – questions like how interactive virtual tours can be, how much 'playfulness' ('Spielelement') is commendable, and in general, what is effective in the dialogue with young people and for preparing them for a visit.[4]

The Dachau Memorial Site will undergo major changes in the next 10 to 15 years. Again, it becomes evident that the memorial landscape for the victims of National Socialistic Germany continues to be fluid and will see new expressions and formulations (Hartmann 2018).

Notes

1 My first visit to the Memorial Site was in 1973, with fellow students at the occasion of a reunion of an Israel trip. I conducted surveys at the entrance to the museum in 1975 (Hartmann 1976). I had meetings with the directors of the Memorial Site, one with Ruth Yakusch, eight to ten meetings with Barbara Distel over 20 years, two meetings with

Gabrielle Hammermann. I had several meetings with members of the Education Department of the Memorial Site: Peter Koch, Waltraud Burger, Stephanie Pilzweger-Steiner and Ulrich Unseld, and most recently, with Kerstin Schwenke on September 2, 2022. I am grateful for the information I received at these meetings. I also had the chance of meeting Harold Marcuse, an outstanding researcher on Dachau, at several occasions as we developed eventually forms of collaboration (Ashworth & Hartmann 2005). I appreciated very much these contacts with him and the information he freely provided.

2 "Ach was, da gehen nur die Amerikaner hin (only the Americans go there/to Dachau)" was the response of a Professor at the Munich University in 1974 when I asked him whether I should study tourism to the Dachau Memorial Site, a major attraction in the Munich Metropolitan Area.

3 I remember a visit to the Dachau Memorial Site in 1986 (or 1987), on a Friday afternoon of a hot summer day, when the small management team of the Memorial Site scrambled to deal with a never ending stream of tourists arriving at the site/the entrance to the museum.

4 Information I received at a meeting on September 2, 2022, with Kerstin Schwenke, new director of the Education Department at the KZ-Dachau Memorial Site.

References

Ashworth, G. & R. Hartmann. (Eds.). (2005). *Horror and Human Tragedy Revisited – The Management of Sites of Atrocities for Tourism.* New York: Cognizant Communication Corporation.

Bavarian Memorial Foundation. (2023). *Anniversary – 20 Years Bavarian Memorial Foundation: Looking Back to the Future – The Bavarian Memorial Foundation Celebrates Its 20th Anniversary.* https://www.stiftung-bayerische-gedenkstaetten.de/en/anniversary

Benz, W. (2008). Gedenkstaettenarbeit in Dachau: Barbara Distel zum Abschied. In W. Benz & A. Koenigseder (Eds.), *Das Konzentrationslager Dachau – Geschichte und Wirkung nationalsozialistischer Repression.* Berlin: Metropol Verlag, 13–16.

Benz, W., B. Distel & A. Koenigseder. (2009). Der Ort des Terrors. Geschichte der nationalsozialistischen Konzentrationslager in neun Baenden. *Dachauer Hefte,* 25, 301–312.

Burger, W. (2017). *Learning – Remembering – Meeting. Educational Work at the Dachau Concentration Camp Memorial Site.* Dachau: KZ-Gedenkstaette Dachau/Stiftung Bayerische Gedenkstaetten.

Dachauer Gaestefuehrer. (2021). *Themenfuehrungen 2021 – Dachau und Umgebung.* Dachau: Stadt Dachau, Abt. Tourism's.

Distel, B. & R. Jakusch. (1978). *Konzentrationslager Dachau 1933–1945.* Dachau: Comite International de Dachau.

Hammermann, G. (2003). Das Internierungs- und Kriegsgefangenenlager Dachau 1945–1948. *Dachauer Hefte,* 19, 48–70.

Hammermann, G. (2021). Die KZ-Gedenkstaette Dachau – Zukunft der Erinnerung. *Geschichte in Wissenschaft und Unterricht,* Jahrgang 72, Heft 3–4, 125–144.

Hammermann, G. & S. Pilzweger-Steiner. (2017). *Dachau Concentration Camp Memorial Site – A Tour.* Dachau: Dachau Concentration Camp Memorial Site.

Hammermann, G., J.-M. Thomas, S. Pilzweger-Steiner, M. Schweitzer, B. Meyer-Krahmer & C. von Wedemeyer. (2020). *Das Konzentrationslager Dachau – Ein Film* (text in German as well as in four other languages). Dachau: KZ-Gedenkstaette Dachau/Stiftung Bayerische Gedenkstaetten, 65 pages.

Hartmann, R. (1976). *Das raeumliche Verhalten junger nordamerikanischer Touristen in Bayern.* Unpublished Master's Thesis. Technische Universitaet Muenchen, Muenchen.

Hartmann, R. (1989). Dachau Revisited: Tourism to the Memorial Site and Museum of the Former Concentration Camp. *Tourism Recreation Research,* 14(1), 41–47 (Reprinted in *Tourism Environment,* Tej Vir Singh, et al. (Eds.), New Delhi Inter Indian Publications, 1992, 183–190).

Hartmann, R. (2004). Zielorte des Holocaust-Tourismus im Wandel – die KZ-Gedenkstaette in Dachau, die Gedenkstaette in Weimar-Buchenwald und das Anne-Frank-Hause in Amsterdam. In C. Becker, H. Hopfinger & A. Steinecke (Eds.), *Geographie der Freizeit und des Tourismus*. Muenchen: Oldenboureg Verlage, 297–308.

Hartmann, R. (2017). Places with a Disconcerting Past: Issues and Trends in Holocaust Tourism. *EuropeNow*, 10. http://www.europenowjournal.org/2017/09/05/places-with-a-disconcerting-past-issues-and-trends-in-holocaust-tourism/

Hartmann, R. (2018). Tourism to Memorial Sites of the Holocaust. In P. Stone, R. Hartmann, T. Seaton, R. Sharpley & L. White (Eds.), *The Palgrave Handbook of Dark Tourism Studies*. London: Palgrave Macmillan, 469–507.

Mannheimer, M. (1985). Theresienstadt – Auschwitz – Warschau – Dachau Erinnerungen. *Dachauer Hefte*, 1, 88–128 (In English translation (2003) "Theresienstadt – Auschwitz – Warsaw – Dachau Recollections". In W. Benz & B. Distel (Eds.), *Dachau and the Nazi Terror, Vol. I, 1933–1945 Testimonies and Memoires*, 9–49).

Mannheimer, M. (2008). Von Auschwitz nach Karlsfeld und Muehldorf. In W. Benz & A. Koenigseder (Eds.), *Das Konzentrationslager Dachau – Geschichte und Wirkung nationalsozialistischer Repression*. Berlin: Metropol Verlag, 450–452.

Mannheimer, M. (2009). Einstieg in die Zeitzeugenarbeit. *Dachauer Hefte*, 25, 324.

Marcuse, H. (1990). Das ehemalige Konzentrationslager Dachau: Der muehevolle Weg zur Gedenkstaette, 1945–1968. *Dachauer Hefte*, 6, 182–205.

Marcuse, H. (2001). *Legacies of Dachau – The Uses and Abuses of a Concentration Camp*. New York: Cambridge University Press.

Marcuse, H. (2005). Reshaping Dachau for Visitors 1933–2000. In G. Ashworth & R. Hartmann (Eds.), *Horror and Human Tragedy Revisited – The Management of Sites of Atrocities for Tourism*. New York: Cognizant Communication Corporation, 118–148.

Marcuse, H. (2010). The Afterlife of the Camps. In J. Kaplan & N. Wachsmann (Eds.), *Concentration Camps in Nazi Germany*. New York: Routledge, 186–211.

Richardi, H.-G. (1983). *Schule der Gewalt – Das Konzentrationslager Dachau 1933–34*. Muenchen: Verlag C. Beck.

Richardi, H.-G., E. Philipp & M. Luecking. (1998). *Dachauer Zeitgeschichtsfuehrer*. Dachau: Stadt Dachau Amt fuer Kultur, Fremdenverkehr und Zeitgeschichte.

Schwenke, K. (2022). *Fuenf Fragen an Dr. Kerstin Schwenke, Leiterin der Bildungsabteilung, Newsletter der KZ-Gedenkstaette im Maerz*, Maerz 15, info@news.kz-gedenkstaette-dachau.de, pp. 7–8.

Zamecnik, S. (2004). *That Was Dachau 1933–1945*. Paris: Le Cherche Midi.

Zum Beispiel Dachau. (1983). *Die Stadt und das Lager, Arbeitsgemeinschaft zur Erforschung der Dachauer Zeitgeschichte*. Dachau: Exhibition Catalogue.

3 The long and twisted road to a memorial

The Kaufering satellite camp complex of the Dachau Concentration Camp and the difficulties of coming to terms with the past

Manfred Deiler and Edith Raim

3.1 Historical background

The Dachau Concentration Camp had 140 satellite or subsidiary camps (46 *Aussenlager* and 94 *Aussenkommandos*), with most of them established in the time period 1943 to 1945. From June 1944 onward, 11 "Kaufering" camps were situated in the surroundings of the City of Landsberg am Lech, and the neighboring community of Kaufering in Upper Bavaria. As the destination of the trains transporting the concentration camp prisoners was the Kaufering railway station – where nearby also the first camp would be established – all the 11 camps became eventually known under the name of Kaufering (distinguished by Roman numbers). Administratively the 11 Kaufering camps were under the responsibility of the main concentration camp in Dachau.

In the last year of the Second World War, 1944/1945, about 23,500 Jews from all over Europe were deported to the Kaufering camps to construct bunkers, as bomb-proof facilities for the German aircraft industry which by then had been badly damaged by Allied air raids. The majority of the Kaufering camp inmates were men, but there were 4,200 women and 850 children, too. The women's section of the camps was situated within the men's camps but separated by barbed wire. Three very large construction sites came into being, although work continued only on one of them until the end of the war. The partially subterranean bunkers were destined for manufacturing Messerschmitt (Me) 262 fighter planes which were deemed crucial for winning back German air supremacy and a successful continuation of the war. Living and working conditions for the deported Jews were atrocious and led to illness and death. Furthermore, the arbitrariness of the SS functionaries in their punishment of misdemeanors as well as executions decimated the numbers of the inmates. During the ten months of June 1944 to April 1945, about 6,500 victims (known by name) died on site of the Kaufering camps and were buried in mass graves near the camps and the construction sites. Other prisoners who could no longer work were deported from Kaufering to a certain death in Auschwitz-Birkenau or Bergen-Belsen, where a typhoid epidemic raged. And yet, among the peculiarities of the Kaufering camps is the fact that seven children

DOI: 10.4324/9780367823795-6

Figure 3.1 Five of the seven mothers with their babies born in Kaufering Camp I after liberation in May 1945.

Source: Credit: National Archives and Records Administration

were born between December 1944 and February 1945 in camp Kaufering I as their mothers had managed to pass through selection in Auschwitz-Birkenau with their pregnancies unnoticed.

At the end of April 1945, the Kaufering camps were liberated. Again, many concentration camp prisoners lost their lives during the last chaotic days of the war, as the prisoners were forced to evacuate the camps, in "death marches", and on trains mistaken for troop transports bombed by Allied aircraft. One of the camps, Kaufering IV, a sick camp, was burned down.

After the war, the liberated Jews either returned to their home countries or remained temporarily in Displaced Persons camps in Southern Bavaria, to await emigration overseas, to the United States, Canada or South Africa or the nascent state of Israel. The three predominant Displaced Persons camps in this region were Foehrenwald, Feldafing and Landsberg.[1] The mass graves of the Kaufering camp prisoners were turned into cemeteries with memorials; although not all the locations of the mass graves were known, no less than 14 cemeteries came into being in the vicinity of the former camps or near the infirmaries where Jews had been hospitalized after liberation. Jewish survivors continued to visit the cemeteries until the early 1950s. By then, most of them had settled elsewhere. Only very few remained in the region of Upper Bavaria or Swabia. Despite the cemeteries as blatant reminders of mass murder, the local Germans were glad to withdraw into "normality" which mainly meant forgetting about the camps which had been situated near their Landsberg hometown and nearby villages for about ten months. Buildings of the camps had been destroyed at the end of the war either by the SS

Figure 3.2 Camp Kaufering IV shortly after liberation. The photograph taken by American liberators shows the extent of the camp and the arson committed by the SS.

Source: Credit: National Archives and Records Administration

when they fled the scene of their crimes or by American troops in their effort to control diseases.

The remnants of the camps had been either burned or torn apart, as Jewish survivors as well as local Germans picked the last useful items such as wood or nails from the debris and used them for other purposes. By the mid-1950s, virtually every trace of the camps had been erased. Still, however, the nearly finished subterranean bunker (and two unfinished construction sites) hovered around. The newly established *Bundeswehr* (German armed forces) moved into the barracks with the bunker and continue to use it until today. The unfinished construction sites turned into gravel pits. All seemed to point to a quick for-getting of an unpleasant period for Germans. But the remnants of one camp, Kaufering VII, had escaped the general erasure of history due to its peculiar construction. While most of the Kaufering camps sported so-called earthen-huts (basically tents made of plywood and covered with earth) which were quickly put up and equally quickly dismantled, Kaufering VII had been partially built with clay tubes forming a vault, which resulted in a round and surprisingly stable construction.

These huts had housed the female inmates of camp Kaufering VII. But even this last remnant was endangered: in 1979, the site was considered a dangerous hazard by the community of Erpfting where the former camp was located. Only due to the farmer who refused to pay about 1,000 German Marks (DM) for the removal of the buildings, the huts remained.

For nearly 40 years, history seemed to be forgotten. In the mid-1970s, youth organizations of German trade unions visited the concentration camp cemeteries for

Figure 3.3 The construction consisting of clay tubes at Kaufering VII seen from inside.
Source: Credit: Collection Manfred Deiler

commemorations of pogrom night in November. Arguably, this was a "wrong" day as it refers to the plight of German, Austrian and Czech Jews during *Kristallnacht*, the pogrom of November 1938, but it raised awareness of persecution and mass murder of European Jews during the Second World War. The mass grave cemeteries which had largely been left untended were from now on better looked after by the *Bayerische Schlösser- und Seenverwaltung* (Bavarian Palaces and Lakes Administration). Though, knowledge about the origins or the history of the camps did not exist and regular commemorations by the City of Landsberg or its neighboring communities did not take place. A history essay competition was to change this.

3.2 Coping with the past on a local level

In 1982/1983, a nation-wide history competition on the everyday life conditions in the National Socialist era (*Alltag im Nationalsozialismus*) caught the interest of three Landsberg students. They decided to conduct research on the background of the cemeteries which were located in Landsberg County. When they went to the Dachau memorial library and archives for information, they found out that there was little known or documented information. Thus, they started to conduct interviews with inhabitants of Landsberg, including one survivor of the Kaufering camps; their research efforts generated important information which was further pursued in other archives and libraries. The final essay comprised around 300 pages and was awarded the first prize of the German Federal President (*Bundespraesident*).

Both the students and their teacher felt that work on this topic needed to continue. They thought that more research needed to be done in particular with survivors who were living abroad in Israel or the United States as well as research in other important archives which were located in foreign countries. One student wanted to ground future work within the local Landsberg society with further publications. So by November 1983, an association named *Buergervereinigung Landsberg im 20. Jahrhundert* (Citizens' association Landsberg in the 20th century) was founded; it was dedicated to delve into Landsberg's hitherto neglected contemporary history, that is, the local events of the years of the Weimar Republic, the Third Reich and the immediate post-war period. A central project was the preservation of the last remnants of camp Kaufering VII, which, at that time, was still in private hands. During the Nazi dictatorship, lands had often simply been confiscated by the Nazi authorities; in the post-war years it was returned to the rightful owners. Also, it was projected to have a documentation center near the historic camp site.[2]

One important step toward preservation was the application for inclusion of the buildings on a list of protected monuments and landmarks. In 1986, the Bavarian Administration for Monument Preservation complied with this request and certified the buildings and structures of Kaufering VII as official monuments. This meant that no longer could anybody do away with the remnants of the buildings. Another consequence of the act of declaring the structures as monuments was that the farmers who owned the land lost interest in the plot since the protected buildings would prevent them from making agricultural profits. In 1986, with the most generous donation of Alexander Moksel, a Polish-Jewish survivor who lived in the neighborhood of Landsberg, the association *Buergervereinigung Landsberg im 20. Jahrhundert* was able to purchase the land of the former camp where the huts made of clay tubes were located. Two-thirds of the land which originally was part of the camp Kaufering VII was and is owned by the municipality of Landsberg. Now, members of the association could start clearing the lands and the huts of heaps of rubbish which had accumulated during the past decades.

In the following years, important steps were taken: a scholarly article was published[3] and a dissertation[4] written. Survivors recounted their experiences in oral history interviews which were recorded or filmed, and books with their memoirs or articles on their experiences were published.[5] The association *Buergervereinigung Landsberg im 20. Jahrhundert* regularly organized commemorations on the site of the former camp Kaufering VII and had a memorial established at the site of Kaufering III. In 1984, for the inauguration, former prisoner the world-famous neurologist Victor E. Frankl[6] came and gave a moving account of his concentration camp imprisonment at Kaufering III and Kaufering VI.

Survivors from other European countries, Israel and the United States started to take a greater interest in what was happening at the site of their former suffering. The association also published a series of booklets titled *Themenhefte*[7] to further knowledge about Landsberg's contemporary history among the general public. Guided tours for school classes as well as for the general public took place on a regular basis and continue to do so until today. The same holds true for research as many documents are still to be unearthed. Over the years, contact with survivors or their relatives became the single-most important work done by volunteer members of the association. The German armed forces (*Bundeswehr*), which used the bunker

Figure 3.4 The photo shows a commemoration in 1984 with the Kaufering survivor Profes-
           sor Viktor E. Frankl.
Source: Credit: Collection Manfred Deiler

codenamed *Weingut II*, learned from the good example set by the association and
established a memorial for the victims as well as an exhibition on the former construction site, which is regularly updated.

By the early 1990s, a new threat arose, this time coming from Landsberg's city
council, which planned a road that would touch and indeed cross the plot of land
formerly forming Kaufering camp VII. In order to at least symbolically protect
the territory, the association *Buergervereinigung Landsberg im 20. Jahrhundert*
decided to ask European heads of state to donate memorial stones which would
be mounted on the site. They represented the countries whose nationals had been
incarcerated and died in the satellite concentration camp complex Kaufering. Over
the years, nearly all the heads of state donated memorial stones which are situated
at the site of the former camp. US army veterans dedicated a memorial plaque to
commemorate the contribution of the liberators and were present at the inauguration ceremony. The reasoning of the association helped. The planned road now
circumvented the former territory of the camp, and no further diplomatic problems
arose as the memorials of the heads of state did not have to be re-located.

In 2009, the association received again a substantial donation from a survivor of
Kaufering, with the request of encouraging the establishment of a foundation which
would expand the legal grounding of the association. So the European Holocaust
Memorial Foundation (*Europaeische Holocaust-Gedenkstaette Stiftung*) came into
being, with the consequence of becoming the owner of the plot of land holding the
former concentration camp buildings.

One ever-more pressing problem was the deterioration of the buildings on
the site of Kaufering VII after the structures had braced the inclement conditions

Figure 3.5 One of the memorial stones at the former camp Kaufering VII commemorates
the contribution of the American troops as liberators of the camps.

Source: Credit: Collection Manfred Deiler

for more than six decades. In order to protect them from vandalism, a fence was
erected, which surrounded the plot of land as well as the buildings. Though, this
was only a feeble and temporary measure. In 2009, the European Holocaust Memorial Foundation formally approached the Bavarian Administration for Monument
Preservation, which had placed the buildings under monument protection in the
mid-1980s. The foundation asked for probing the possibility of conservation measures for the buildings. The legal situation was highly complicated. As successor
state, the Federal Republic of Germany "inherited" the remaining buildings of the
former concentration camp Kaufering VII erected during the Nazi rule. The Federal Republic of Germany claimed it had only safeguarding duties for the buildings
in order to prevent accidents or damage but excluding any claims for compensation
for such damages. Thus, it would not include preservation or conservation measures for the structure of the buildings. While the federal authorities were in charge

Figure 3.6 The preserved buildings of the former camp Kaufering VII after the conservation measures 2014–2016.

Source: Credit: Collection Manfred Deiler

of the buildings, the European Holocaust Memorial Foundation owned the land on which the buildings stand.

Negotiations between the Federal Real Estate Administration, the authority on the Federal side, and the European Holocaust Memorial Foundation led to the decision of a threefold process. First, the Bavarian Administration for Monument Preservation would undertake a feasibility study to probe into the possibilities and problems of conservation measures. A thorough analysis initiated in 2012 established that conservation measures were possible and should be pursued. Second, the Federal Real Estate Administration ceded the buildings to the European Holocaust Memorial Foundation on January 1, 2013. Third, the liability insurance for this extraordinary project of conservation measure for a former concentration camp with unique and peculiar construction features could be obtained and also started on January 1, 2013.

In 2014, conservation measures were started, which continued until the end of the year 2016. Several partners, including the feds, the state, regional authorities and the non-profit foundation, shared in the costs for the conservation of the buildings. Though, the partnership did not include the City of Landsberg; they neither contributed nor supported the project. The completed project of preserving the buildings of the former camp Kaufering VII was awarded several prizes, among them the gold medal award for Bavarian Monument Preservation in 2016.

3.3 The main project: a documentation center

After the successful completion of the conservation measures at camp Kaufering VII, a new project was taken: a documentation center and memorial site for the Kaufering satellite camp complex. The process that eventually led to a comprehensive conceptualization of a documentation center in the Kaufering VII area was highly complex. It was largely initiated by the non-profit European Holocaust Memorial Foundation, with the support of the Bavarian Administration for Monument Preservation and the Bavarian Council for Monument Preservation. The ideas generated in the process were presented to the Bavarian Memorial Foundation (*Stiftung Bayerische Gedenkstaetten*), the agency in charge of the two main concentration camps in Bavaria, Dachau and Flossenbuerg, 75 concentration camp cemeteries and other places of remembrance (*Erinnerungsorte*) regarding the crimes of National Socialism. In 2019, negotiations of a steering committee, with members of the Bavarian Ministry of Culture and Education, the Bavarian Memorial Foundation and the European Holocaust Memorial Foundation, resulted in the recommendation for a scholarly concept concerning the future development of the memorial site Kaufering VII, which was completed in 2022.[8]

Unfortunately, on the local level little progress was made. A feasibility study financed by the Bavarian Memorial Foundation saw the formation of an acting group (among members of the European Holocaust Memorial Foundation, of the City of Landsberg and of the Federal Armed Forces as the bunker was located at their garrison in Landsberg) that fell apart without a tangible result. Thus, the documentation center could not get the OK and go-ahead from the Landsberg City Council, which had ownership responsibilities for two-thirds of the land at the proposed site.

What are the main goals and objectives of the planned documentation center, which will house an exhibition, a library and archives, with documents from the collections and photographs of the European Holocaust Memorial Foundation? The goal was to present a comprehensive conceptualization of the planned documentation center so that an application for the financial support of the project could be turned in. The establishment of the documentation center would be supported by federal authorities and by the Bavarian Ministry of Culture and Education. The author of the scholarly concept argued that the site at Kaufering VII was of national importance as the buildings are the only originally preserved historic prisoners' barracks in Germany, that it showed the European dimension of the mass murder of the Jewry and that it told the story of the racist persecution of children and juveniles. At the center of the planned documentation center was the epoch-forming year of 1945, drawing together three narratives: that of the life and end of persecuted Jewish communities in the East, the work and living conditions in the Kaufering satellite camp complex, and the immediate post-war situation of the liberated prisoners in the Displaced Persons' camp in Landsberg. Other rarely told parts of the history are acts of resistance as well as cultural and religious practices in the Kaufering camps. Further considerations for the exhibit are a focus on the role of the perpetrators and the involved industrial enterprises, with a prevailing agenda of 'extermination through work' (*Vernichtung durch Arbeit*). The Bavarian Ministry of Culture and Education will take steps to institutionalize the staff's work at the center, to finance the exhibition and a building which includes seminar rooms for pedagogic work.[9]

While the proponents of the planned documentation center find themselves currently in a complex decision-making situation, other projects continue to be carried out. The School of Applied Sciences in Augsburg as well as the Technical University in Munich had several students' projects involving the history of the Kaufering camps. The huge collection of documents, photographs and data continues to be used by researchers and students alike. Consistent support for the work of the European Holocaust Memorial Foundation comes also from a local Landsberg member of the Bavarian Parliament, with participation in arranged committee meetings and statements for the press and the public. Two Israeli consulate generals, based in Munich, have taken an interest in the development of the memorial site and visited the remnants. Site visitations by school groups often led by specially trained class mates continue to date. Another outstanding project is the professional production of plaques at camp Kaufering IV (Hurlach) and nearby cemeteries.

In retrospect, we can identify three retarding factors in this long and twisted road toward a memorial site. First of all, research was lagging behind. For a considerable span of time, well into the 1990s, the important role of the satellite or subsidiary camps in Germany and Eastern Europe was hardly acknowledged. Marcuse, who reviewed the afterlife of the concentration camp system, saw a distinct neglect of scholarly work on the subsidiary camps.[10] Schalm, who presented a dissertation on the subsidiary camps of Dachau, discussed in detail the initial lack of research and the low regard or disregard of oral history documents as provided by many prisoners of the satellite camps.[11] While the history of the main camps became common knowledge because of the frequent book and article publications, the outer camps were considered a less important appendix to the main narrative of violence in the main camps. Further, the main concentration camps such as Dachau became a memorial, with an exhibit and other documentations on site. The subsidiary camps' crucial role in the armament industry was hardly researched at all. This started to change in the late 1980s and the early 1990s and continues until today, as many facts are still unknown. The memoirs of survivors of the satellite camps were also neglected for a long time. Second, an overarching umbrella organization responsible for the concentration camps and their memorials did not exist until 2003. Then, the Bavarian Memorial Foundation was set up, which is meanwhile responsible not only for the two main memorials Dachau and Flossenbuerg in the State of Bavaria but also for many other places tied to the crimes during the national socialistic era now with regular commemoration practices for the victims. Third, the municipality of Landsberg and neighboring communities such as Kaufering were for a long time anxious not to be associated with the history of the camps. Places like Dachau had a "stigma" better to avoid.[12] Thus, some of the mayors felt it was more prudent to belittle the history of the nearby satellite camps or to deny their existence. As time passed, the citizens who lived close to the camps during the Nazi dictatorship are no longer alive and painful local memories may have subsided. New generations and younger people feel less inhibited to deal with the drama of the past and some of them have even developed an interest or curiosity in the events that happened once in their communities.

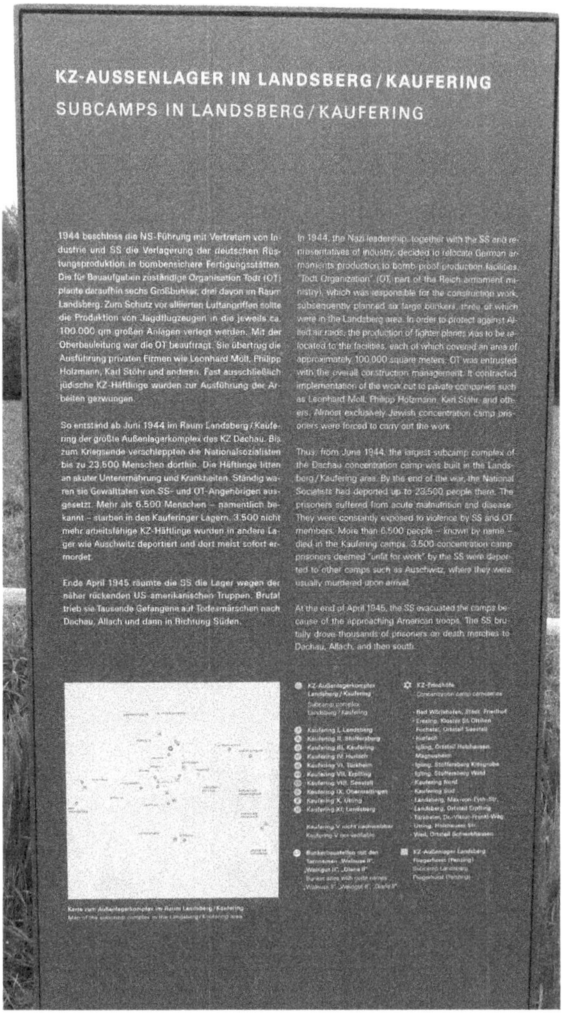

KZ-AUSSENLAGER IN LANDSBERG/KAUFERING
SUBCAMPS IN LANDSBERG/KAUFERING

1944 beschloss die NS-Führung mit Vertretern von Industrie und SS die Verlagerung der deutschen Rüstungsproduktion in bombensichere Fertigungsstätten. Die für die Bauaufgaben zuständige Organisation Todt (OT) plante daraufhin sechs Großbunker, drei davon im Raum Landsberg. Zum Schutz vor alliierten Luftangriffen sollte die Produktion von Jagdflugzeugen in die jeweils ca. 100.000 qm großen Anlagen verlegt werden. Mit der Oberbauleitung war die OT beauftragt. Sie übertrug die Ausführung privaten Firmen wie Leonhard Moll, Philipp Holzmann, Karl Stöhr und anderen. Fast ausschließlich jüdische KZ-Häftlinge wurden zur Ausführung der Arbeiten gezwungen.

So entstand ab Juni 1944 im Raum Landsberg/Kaufering der größte Außenlagerkomplex des KZ Dachau. Bis zum Kriegsende verschleppten die Nationalsozialisten bis zu 23.500 Menschen dorthin. Die Häftlinge litten an akuter Unterernährung und Krankheiten. Ständig waren sie Gewalttaten von SS- und OT-Angehörigen ausgesetzt. Mehr als 6.500 Menschen – namentlich bekannt – starben in den Kauferinger Lagern. 3.500 nicht mehr arbeitsfähige KZ-Häftlinge wurden in andere Lager wie Auschwitz deportiert und dort meist sofort ermordet.

Ende April 1945 räumte die SS die Lager wegen der näher rückenden US-amerikanischen Truppen. Brutal trieb sie Tausende Gefangene auf Todesmärschen nach Dachau, Allach und dann in Richtung Süden.

In 1944, the Nazi leadership, together with the SS and representatives of industry, decided to relocate German armaments production to bomb-proof production facilities. "Todt Organization" (OT), part of the Reich armament ministry, which was responsible for the construction work, subsequently planned six large bunkers, three of which were in the Landsberg area. In order to protect against Allied air raids, the production of fighter planes was to be located to the facilities, each of which covered an area of approximately 100,000 square meters. OT was entrusted with the overall construction management. It contracted implementation of the work out to private companies such as Leonhard Moll, Philipp Holzmann, Karl Stöhr and others. Almost exclusively Jewish concentration camp prisoners were forced to carry out the work.

Thus, from June 1944, the largest subcamp complex of the Dachau concentration camp was built in the Landsberg/Kaufering area. By the end of the war, the National Socialists had deported up to 23,500 people there. The prisoners suffered from acute malnutrition and disease. They were constantly exposed to violence by SS and OT members. More than 6,500 people – known by name – died in the Kaufering camps. 3,500 concentration camp prisoners deemed "unfit for work" by the SS were deported to other camps such as Auschwitz, where they were usually murdered upon arrival.

At the end of April 1945, the SS evacuated the camps because of the approaching American troops. The SS brutally drove thousands of prisoners on death marches to Dachau, Allach, and then south.

Figure 3.7 Kaufering satellite camps: overview information.

Source: Credit: Bavarian Memorial Foundation

3.4 Conclusion and outlook

For the last 40 years, responsibility for the site of the former concentration camp Kaufering VII rested on the shoulders of volunteers gathered first in the *Buergervereinigung* association, then in the European Holocaust Memorial Foundation. With the creation of the European Holocaust Memorial Foundation, research has been more professionalized, with high-quality documentations about the Kaufering satellite camp system. The foundation reached out to survivors and the local public and organized and supervised the conservation of the original huts in cooperation

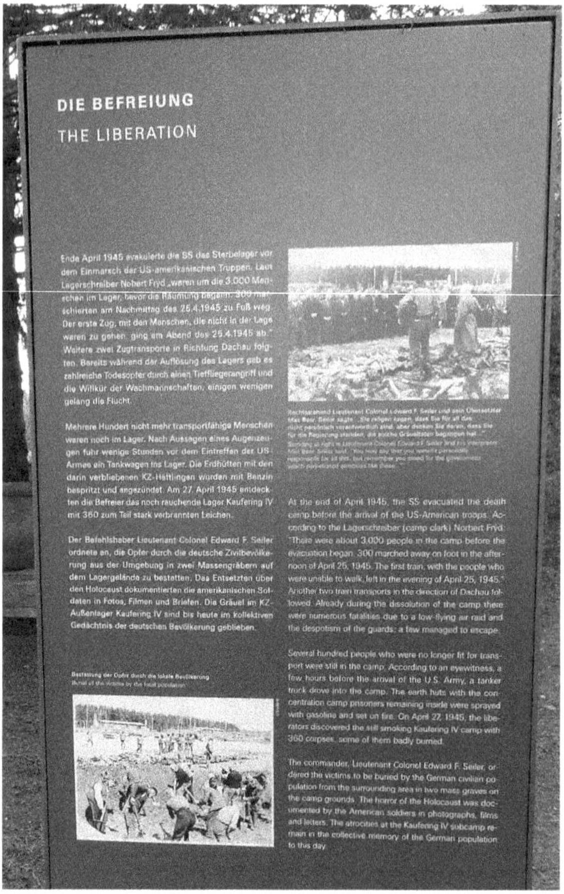

Figure 3.8 Kaufering IV Camp. Site-specific information. The plaques inform about the history of the Kaufering camp number IV near Hurlach and its gruesome end as it was burned down in late April 1945.

Source: Credit: Bavarian Memorial Foundation

with Federal and Bavarian authorities. Over several decades, they held innumerable commemorations and created memorials not only at camp Kaufering VII but at other Kaufering camps and cemeteries as well. Recently, information plaques have gone up (in collaboration with the Bavarian Memorial Foundation) at the former camp site Kaufering VIII and Kaufering IV; others are being planned.

In our view, future developments are beyond the scope of voluntary work. Visits to the site of the former camp Kaufering VII are hindered by many factors. As there is no visitor center at the site and no shelters, visitors are exposed to the elements. There are no seminar rooms where groups could discuss their experiences and exchange opinions, not to mention the lack of sanitary facilities. Although interest in the history of the Holocaust and the Second World War is still on the rise, increasing demand is difficult to handle. More and more enquiries for tours come from school classes, which can be accepted only to a smaller degree as the voluntary guides are limited to small groups of visitors on their walk through the former camp site. And groups who can secure a slot for a visit have to cope with the adverse conditions as described earlier.

The European Holocaust Memorial Foundation is devoted to the principles of the International Memorial Museums Charter of the International Holocaust Remembrance Alliance (IHRA), which was adopted in 2012 and amended in 2016.[13] In the year 2000, representatives from 46 countries, among them 23 heads of states and 14 deputy heads, signed the Stockholm declaration. They committed themselves to further the remembrance of the Holocaust in education, commemoration and scholarship. They stressed the universal meaning of the Holocaust and the permanent task of commemoration. The new memorial site at the former camp Kaufering VII has to comply with the charter of internationally agreed-upon principles and ethics.[14]

As to the future of the documentation center and memorial site for the Kaufering satellite camps, there is a clear need for a support system which should be provided by the Federal Republic of Germany and the State of Bavaria. The successful management and maintenance of the documentation center must rest on sufficient funding. The planned visitor center, for instance, with seminar rooms, a library and archives, needs permanent staff. The guides who take the visitors around the site have to be well trained, at the Dachau memorial site where courses for accreditation are regularly conducted and at Kaufering. Publication and internet activities, including social media contributions, have to be continued. Keeping up the archival work and the collections as well as the correspondence with survivors and their relatives is another part of the agenda of the planned documentation center. Commemoration events with speakers at the end of April (anniversary of the liberation of the camps) have to be organized.

While volunteer work for a transitional period at the site is certainly useful, if not necessary, permanent solutions need to be sought and found. One cannot overrate the importance of this project in a time when the erosion of democracy has become a real threat in Europe. Remembrance at the camp Kaufering VII site began with volunteer organizations. It is proof of the significance of a vibrant democratic civil society, reminding the state of its duties toward remembrance and commemoration of the persecuted.

Notes

1 For a general overview, see Atina Grossmann, *Jews, Germans, and Allies: Close Encounters in Occupied Germany* (Princeton: Princeton University Press, 2009).

2 For more information, see http://www.landsberger-zeitgeschichte.de/Gedenkstaette.htm [Last access: Oct 5, 2023].

3 Edith Raim, Unternehmen Ringeltaube & Dachaus Aussenlagerkomplex Kaufering, *Dachauer Hefte: Die vergessenen Lager*, vol. 5 (Munich: DTV, 1994), pages 193–213.

4 Edith Raim, *Die Dachauer KZ-Aussenkommandos Kaufering und Muehldorf. Rustungs-bauten und Zwangsarbeit im letzten Kriegsjahr 1944/45* (Landsberg: Landsberger Verlagsanstalt Rudolf Neumeyer, 1992).
5 *Dachauer Hefte*, passim.
6 Viktor E. Frankl, *Yes to Life in Spite of Everything* (London: Rider, 2021); Viktor E. Frankl, *Man's Search for Meaning* (New York: Random House, 2021).
7 Themenheft 1: *Von Hitlers Festungshaft zum Kriegsverbrecher-Gefaengnis No. 1: Die Landsberger Haftanstalt im Spiegel der Geschichte*; Themenheft 2: *Todesmarsch und Befreiung – Landsberg im April 1945: Das Ende des Holocaust in Bayern*; Themenheft 3: *"Der nationalsozialistische Wallfahrtsort Landsberg 1933–1937: Die "Hitlerstadt" wird zur "Stadt der Jugend"*; Themenheft 4: *Das KZ-Kommando Kaufering 1944/45: Die Vernichtung der Juden im Ruestungsprojekt "Ringeltaube"*; Themenheft 5: *Sonderheft 50 Jahre Befreiing. Das SS-Arbeitslager Landsberg 1944/45: Franzoesische Widerstandskaempfer im deutschen KZ*; Themenheft 6: *Landsberg 1945–1950: Der juedische Neubeginn nach der Shoa. Vom DP-Lager Landsberg ging die Zukunft aus* (Landsberg: Buergervereinigung, 1993–1996).
8 E. Raim, *Wissenschaftliches Konzept fuer die Errichtung eines Lern-, Erinnerungs- und Gedenkorts am ehemaligen KZ-Aussenlager Kaufering* (Landsberg: Europaeische Holocaustgedenkstaette Stiftung e.V., 2022), unpublished document, 52 pages.
9 E. Raim, *Wissenschaftliches Konzept fuer die Errichtung eines Lern-, Erinnerungs- und Gedenkorts am ehemaligen KZ-Aussenlager Kaufering* (Landsberg: Europaeische Holocaustgedenkstaette Stiftung e.V., 2022), unpublished document, 52 pages.
10 H. Marcuse, The Afterlife of the Camps. In J. Kaplan & N. Wachsmann (Eds.), *Concentration Camps in Nazi Germany* (New York: Routledge, 2010), pages 186–211.
11 S. Schalm, Ueberleben durch Arbeit? Aussenkommandos and Aussenlager des KZ Dachau 1933–1945. In *Geschichte der Konzentrationslager 1933–1945 Band 10* (Berlin: Metropol Verlag, 2009), pages 11–24.
12 G. Hammermann, Die KZ-Gedenkstätte Dachau – Zukunft der Erinnerung. *Geschichte in Wissenschaft und Unterricht*, 72, no. 3–4 (2021), pages 125–144.
13 https://www.holocaustremembrance.com/de/about-us/stockholm-declaration.
14 https://www.holocaustremembrance.com/resources/working-definitions-charters/international-memorial-museums-charter.

4 German landscapes of commemoration

The difficult legacy of wartime aerospace industries

Dietrich Soyez

4.1 Introduction

4.1.1 General background and objectives of the article

Industrial heritage valorizations ideally include what may be called 'representative excerpts' from our industrial past. Within the growing field of heritage approaches an impressive broadening of conceptual and topical views has occurred during the past few decades, both intra- and interdisciplinary. However, from a heritage perspective inspired by geographical conceptualizations, some delicate blind spots persist:

- while brighter aspects of industrialization remain at the forefront, the role of uncomfortable historic events is silently passed over, played down or covered up
- the main focus is on individual buildings or factories whereas little attention is paid to functional links between industrial systems, in particular in multi-tiered and multi-local supply networks
- industrial heritage sites remain markedly 'national' in terms of the politics of their designation and interpretational thrust, often neglecting one of the industrial world's constitutive facts: Its transnational patterns of interaction

Hence, it is worthwhile to focus on the German aerospace industry of late WWII as an industrial sector. It has never been addressed more systematically from a heritage perspective but permits scrutiny of all blind spots mentioned earlier. Due to this period's particular challenges and turmoil, the evolution of the German aerospace industry's production systems was strongly influenced by leaps in technology and communications, radical shifts in the labor force and successive disruptions. Furthermore, this situation was constantly superimposed by processes of accelerated outsourcing and boundary-transcending forms of corporate organization and production.[1]

The above issues may appear too complex to be dealt with in a short article. Yet, the use of the two terms *rationalization* and *commodification of misery* permits a revealing examination of both internal processes and external impacts of the evolution of the German aerospace industry during the later stages of the war and resultant heritage issues.

DOI: 10.4324/9780367823795-7

Rationalization pervasively marks this industrial sector's late-war technical development. It mirrors not only the increasing internal refinement, growth, differentiation and efficiency of production system. In this specific case, it also had much wider and shameful implications, not only domestically but beyond Germany's borders, through the systematic *commodification of misery*. This term is understood as an appropriate metaphor for the industry's systematic utilization of forced labor, be they originally formally contracted German and foreign civil workers, prisoners of war or detainees of extermination and other concentration camps (many of them foreign citizens).[2] [3] [4] [5] [6]

Against this background, the article's main objectives are, first, to describe the links between the aerospace industry, its rationalization and forced labor and, second, to review the representation of these issues in today's landscapes of commemoration.

4.1.2 *Main issues at stake: a brief literature review*

The following reflections are situated at the intersection of contemporary industrial heritage approaches and dark (heritage) tourism and transnationalities, partly linked to Holocaust issues (Ashworth & Hartmann 2005; Graham, Ashworth & Tunbridge 2000; Hartmann 2013, 2017a, 2017b; Logan & Reeves 2009; Steinecke 2021). These references constitute an appropriate backdrop for our studies. Though, the more specific questions that will be addressed in the following, that is, the intricate links between air armament and forced labor, have never been discussed systematically in the increasingly interdisciplinary field of industrial heritage (Soyez 2009; Li & Soyez 2016).

This remarkable gap in the literature can no longer be blamed on a lack of data. Academic discourse has gone well beyond its early and almost exclusive focus on extermination camps, such as Auschwitz/Oświęcim or Treblinka. An increasing abundance of documentation is now accessible in historical treatises. They include more consistently the links between the arms industry on the one hand and the long-neglected subsidiary camp system with its forced labor workforce, on the other. It can be stated as fact that already during the early years of WWII sites of industry and forced labor camps were mutually constitutive, even in an infamous camp such as Auschwitz/Oświęcim (cf. Jeffreys 2010). As stated clearly by Steinbacher (2005:3) regarding Auschwitz: "The connection between the intention to exterminate and industrial exploitation became an immediate reality here".

Examples of these linkages are increasingly found also in both company and technical histories (Pohl, Habeth & Brüninghaus 1986; Hamburger Stiftung für Sozialgeschichte des 20. Jahrhunderts 1987; Hopmann et al. 1994; Gregor 1998 on Daimler-Benz; Mommsen & Grieger 1996; Volkswagen AG/Historische Kommunikation 1999 on Volkswagen; Werner 2006 on BMW; Kukowski & Boch 2014 on Auto Union (today Audi), and Overy 1980; Braun 1990; Budraß & Grieger 1993; Budraß 1998, respectively, on broader technological issues). Further, relevant information for our studies is available from Holocaust-related literature (Raim 1992, 1998; Allen 2002; Steinbacher 2005; Wagner 2004a, 2009; Benz & Distel 2006 to 2009 (altogether nine volumes), Megargee 2009; Schafft & Zeidler 2011 and, with regard to the terrible Holocaust aftermath until the mid-1950s for millions of Displaced Persons, Jähner 2021).

The low visibility of 'dark', 'uncomfortable' facts in the industrial heritage literature, however, seems to be a consequence of specific post-war political, academic, corporate and military perceptions as well as dominant discourses, reflecting long prevailing power-structures in Germany. They have led to suppressed, covered-up or forgotten memories. It may also be connected to a feared stain on the reputation of former industrial icons (both entrepreneurs and brands) and their remarkable post-war continuities, many of them revered to the present day (see Budraß 2016a, 2016b and his deconstruction of the post-war iconization of personalities, products and brands of the WWII air industry). The post-war narratives – and myths – with regard to the development and construction of the ballistic rocket A4 at Peenemünde represent another case for post WW II improper de-contextualizing the situation there eventually leading to silenced and repressed feelings (see Section 4.4.1 of Chapter 4).

Thus, while general data regarding central actors in the arms industry including the aerospace sector are now widely available, a very specific reality of late-war conditions can only be partially tracked in the references mentioned. This is particularly true for the complex interactions between main 'production' sites (often just assembly plants), branch plants, multi-tiered supplier networks, on the one hand, and their spatial impacts, on the other, both nationally and beyond Germany's international borders.

Only a better understanding of the functional inner workings of these systems as well as their transnational spatial reach and impact will complement the overall picture. Addressed as 'functionalities' and 'spatialities' in what follows, they will constitute a more reliable basis for appropriate criteria for potential complements of existing landscapes of memory.

4.1.3 Conceptual remarks

The massive transboundary spread and diversification of industrial production systems along with a number of related issues (such as new forms of labor abuse) are at the center of interest of Economic and Industrial Geography since the 1980s. Quite early, they were aptly epitomized by the term 'global shift' (Dicken 1992, 2015). While this term addresses the general dynamics of recent developments, the terms Global Production Networks/Global Value Chains focus in more detail on their functional and spatial configurations (Coe & Yeung 2015; Hayter & Patchell 2016). At this point, the term 'global' does not refer to truly planetary processes, but a big variety of transboundary or supra-national interlinkages, be they chains, networks or circuits. These issues now constitute broad and vibrant fields of research, fortunately not in a context of worldwide disastrous events but (mostly) peaceful transboundary competition and cooperation.

Despite radical contextual differences between peacetime and war economies, a number of these contemporary organizational patterns seem to bear traits comparable to the late-war German aerospace industry's boundary-transcending reach, even if it was almost exclusively limited to those parts of Europe that were under

National Socialist rule. Thus viewed, the late-war evolution of the German aerospace industry is another case in point for a reframing of industrial heritage issues from a geography-inspired perspective.

However, as there are very few truly globalizing, that is, planetary, processes in the world, the above-addressed issues are characterized in the following as transboundary, boundary transcending or transnational and, in the case of tangible or intangible outcomes: Transnationalities.

Some of late-war German aerospace industry's crucial trends, events and impacts have been examined, as outlined earlier, by historians using a variety of approaches. These mirror not only economic, politico-institutional and corporate facts but also decisive tipping points, crises and interrelationships. However, none of these informative and mutually complementary studies aims at putting a special focus on transnational functionalities and spatialities that have not only shaped this industry's remarkable change in continental theaters of war but also left innumerable tangible legacies in impacted landscapes (from machinery to factories and general infrastructure). Yet, all of these historic treatises abound in technical and other details that are illustrative of the specific objectives of this study, and consequently, they will be drawn upon for both concepts and data.

Searching for potential heritage evidence in Nazi Germany's aerospace industry does not reflect an indifference to the destinies of those who became victims of this technical world. On the contrary, it will increase an understanding of their lives and shameful sufferings as constitutive elements of this still quite arcane part of Germany's industrial history.

4.1.4 Line of argumentation

Introductory remarks in Section 4.1 are followed in Section 4.2 by an outline of the political, corporate and military entanglements of the aerospace industry in a highly turbulent time. After commenting briefly on some of the war's decisive turning points in 1942/1943 and their consequences, Section 4.3 will focus on the last two years of the war, characterized by an almost frantic evolution of the crucial issues outlined earlier: The development of the aerospace industries' production system, now increasingly kept functioning by transboundary connections and a widespread system of subsidiary concentration camps providing forced labor. Section 4.4 will consider if and how Germany's existing landscapes of memory should be complemented by carefully chosen examples of the disconcerting industrial world of late WWII and ensuing post-war corporate continuities that last to the present day. Section 4.5 will offer conclusions.

4.2 A tightrope walk: the aerospace industry entangled in corporate, political and military interests

4.2.1 The general context

After the National Socialist seize of power in 1933 the military aircraft industry experienced enormous growth. Budraß (1998:15) refers to what seems like an unbelievable contrast: While a few small German firms reportedly produced a mere 36 planes in 1932, the huge end-of-war aircraft corporations had an output of about 40,000 aircraft in 1944. As a matter of fact, the military aircraft industry had become a leading German high-tech industry by 1940; this is shown in an impressive and constantly growing network of specialized suppliers and subcontractors. Many of them were highly innovative in aerodynamics, electrical engineering, optics, precision mechanics, aluminum and magnesium alloys.

Yet, throughout its history, the military aircraft industry existed in a highly turbulent environment and suffered risks, instabilities and shocks (for broad overviews and crucial insights in Germany's wartime situation, see Tooze 2006), only to mention:

- critical shifts in Nazi Germany's geopolitical situation and rearmament strategies as well as a disturbing sequence of military successes and catastrophes over time;
- pervasive economic and technological disruptions particularly in the aerospace industry;
- and all this impacted directly by inconsistent and erratic political and military decisions, unfolding with never-ending and contradictory patterns of infighting and cooperation within and between the government, the National Socialist party institutions and the industrial corporations

4.2.2 Crucial players: the main aerospace and automotive industries involved

Some of the most important players arming the German air force between 1941 and the end of the war are listed below (only a few war-related relocations will be mentioned in the following). They all possessed late-war branch plants and licensees all over Germany and Axis countries as well as in occupied regions. Detailed listings and descriptions of their main types of planes, flying bombs, jet engines and ballistic rocket engines are also accessible in the relevant international literature (Overy 1980; Vajda & Dancey 1998; Hirschel, Prem & Madelung 2004; Uziel 2012). Important companies were (in brackets: Locations in post-WWII states):

- **Junkers**/Junkers Flugzeug- und Motorenwerke AG (JFM) (Dessau, now a subdivision of the town of Dessau-Roßlau/State of Saxony-Anhalt)
- **Heinkel**/Ernst Heinkel Flugzeugwerke GmbH (as of 1943 EHAG) (Rostock/State of Mecklenburg-Vorpommern)
- **Messerschmitt**/Messerschmitt AG (Augsburg/State of Bavaria)
- **Henschel**/Henschel Flugzeug-Werke (HFW) (Kassel/State of Hesse)
- **Arado**/Arado Flugzeugwerke AG (Rostock/State of Mecklenburg-Vorpommern)
- **Focke-Wulf**/Focke-Wulf Flugzeugbau AG (City State of Bremen)
- **Fieseler**/Fieseler Flugzeugbau (Kassel/State of Hesse)

Finally, a highly crucial research institution must be added to the companies listed earlier even if its tentative production of ballistic rockets in 1943 was interrupted almost immediately by a British air raid and quickly relocated to an underground facility at Nordhausen/State of Thuringia, namely

- **Heeresversuchsanstalt** (HVA)/State-owned Army Research Center (Peenemünde/State of Mecklenburg-Vorpommern)

The German airforce also had a number of research and test institutions (yet without production facilities). The central one was located in the small community of Rechlin/State of Mecklenburg-Vorpommern (see Beauvais et al. 1998).

An overview of the aerospace industry, however, would be incomplete without considering the automotive industry: All German car producers were functionally – and systematically – integrated into the armament industry, in particular with regard to military jeeps, trucks and tanks as well as the licensed production of single parts and more complex aircraft components. Yet, the listed firms' most crucial contributions were air engines, from traditional piston to cutting-edge jet and rocket engines (main locations and products in brackets):

- **Auto Union AG** (Chemnitz, State of Saxony; air engines: Taucha, ditto)
- **BMW**/Bayerische Motorenwerke AG (München; air engines: München-Allach, both State of Bavaria)
- **Daimler-Benz AG** (Stuttgart, State of Baden-Württemberg; air engines: Genshagen, State of Brandenburg; parts of the ballistic rocket A4, Sindelfingen, State of Baden-Württemberg)
- **Henschel**/Henschel-Werke AG (Kassel, State of Hesse; air engines, ditto, complete airplanes at its Berlin-Schönefeld facility, in 1943 temporary an A4 rocket assembly site at subsidiary Rax-Werke, Wiener Neustadt, Austria)
- **Volkswagen** (Fallersleben, during the Nazi regime known as KdF-Stadt, after the war renamed to Wolfsburg, State of Lower Saxony; flying bombs Fi 103; German 'jeep', the *VW Kübelwagen*)
- **Ford**/Ford-Werke GmbH (Köln, State of North Rhine-Westphalia, owned by Ford, Dearborn MI; airplane parts and components)
- **Opel**/Opel AG (Rüsselsheim, State of Hesse, owned by GM, Detroit MI, airplane parts and components)

Only Henschel built complete airplanes, while Volkswagen, for capacity reasons, temporarily took over the production of the Fieseler Fi 103 guided bomb – called 'buzz bomb' or 'doodlebug' by the British – from its developer in Kassel as of mid-1943. Only a few months later Volkswagen lost its position as a lead producer to the SS-administered Mittelwerk, close to Nordhausen/State of Thuringia (for details, see Section 4.3.1.1, below), mainly due to its vulnerability to air raids (Mommsen & Grieger 1996:677–709).

While all these companies adapted to arms production and usually profited from it, it is well documented that a number of them prepared for post-war priorities more or less secretly, for instance by sending crucial employees, machinery and documents to safer locations. Obviously, this fact has contributed to the automotive firms' fast post-war take-off, first with regard to the car production but later some of the former aerospace industries joined their European counterparts in a rapid restructuring process.

4.2.3 Milestone decisions as sequential replacements of institutional formations

Using the concept of 'institutional formations', Budraß (1998) convincingly argued that the air industry has always been torn between two strategic poles, by prioritizing either technological excellence or strict procurement strategies. The first agenda was pushed mainly by entrepreneurial developers, among them Hugo Junkers, Ernst Heinkel and Willy Messerschmitt. Procurement priorities, by contrast, were mainly imposed by dominant political-administrative experts and military officials in charge at the Air Ministry and its subdivisions. Over the course of the war, both priorities changed several times.

The crucial end-of-war climax of these internal battles and shifts concerning air armament priorities, that is, the conflict-laden power play toward a new – and last-ditch – institutional formation, was what Budraß (1998:705 et seq.) calls a radical procurement focus in March 1944. It was principally triggered by the precision of Allied attacks on aerospace industries all over Nazi Germany during the so-called 'Big Week' (February 8 and 15, 1944, Holland 2018) that had led to a massive destruction of aircraft factories and of hundreds of already operational planes on their adjacent airfields. A new ad hoc task force was created, limited to a period of six months, the so-called Jägerstab (*Fighter Staff*, Allen 2002:ch. 6; Uziel 2012:ch. 3). It comprised experts and officials from involved ministries and aerospace industries and was given almost absolute authority to all but discontinue bomber production as well as steer and increase the production of fighters (for details, see Uziel 2012:87, 2014).

The new focus on defense weapons, in particular fighter planes, also led to the late-war country-wide migration of tens of thousands of (mainly forced) laborers from bomber to fighter factories.

4.3 Planning and waging the war: interconnected loops in and impacts on the aerospace production system

4.3.1 Technical rationalization and ensuing workforce commodification

The following reflections are deeply rooted in and linked with the crucial overarching topics of the general development of the war over time, the National Socialistic party and bureaucracy, the Nazi concentration camp system and the aerospace armament industry. However, as underlined earlier, the focus will be on two crucial interconnected strands within this larger field, namely the *rationalization* of aerospace production, on the one hand, and the *commodification of misery*, representing the completely dehumanized exploitation of forced labor, on the other. In order to make it easier to grasp important events, interconnections and outcomes, a simplified synthesis is provided as a contextual background (see Figure 4.1).

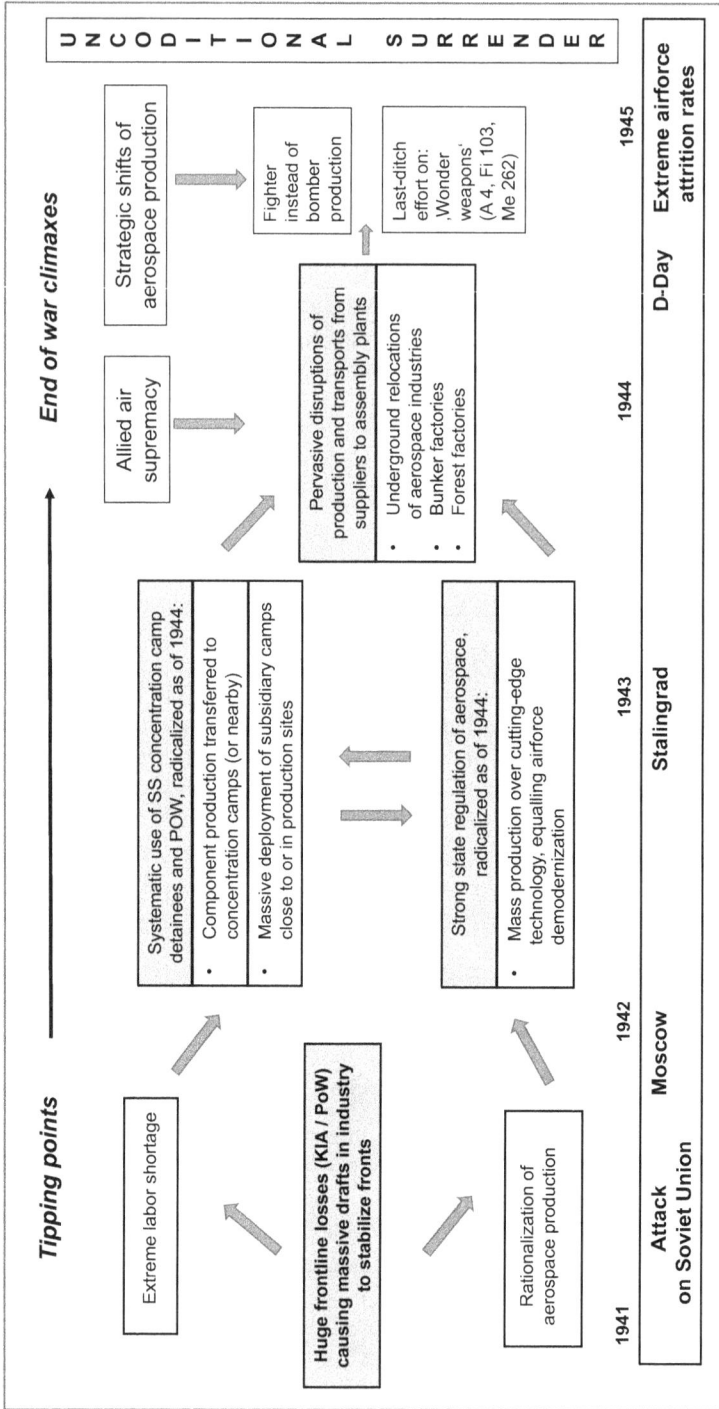

Figure 4.1 German Air Armament as of 1942: stronger state regulation and increased use of forced labor.

The main issue regarding *rationalization* in the aerospace industry can be summarized as developing from workbench- and batch-production to constantly improved flow production. Obviously, the goal was to produce airplanes (almost) as efficiently as mass-produced motor cars by reducing the manual work of skilled workers while introducing high-level standardization, mechanization and single-purpose tool machinery.

However, the technical complexity of aircraft compared to cars was still evident in the 1940s. This applied not only to the number of components to be processed (ten- or even a hundredfold) but also in the much higher precision requirements as well as the aerodynamic requirements of the airframe itself.

Technically, the production process had to be organized as a highly intricate sequence of steps, allowing semi-skilled or even unskilled workers to produce high-tech products with a minimum of manual work. Institutionally, however, this approach devalued both organizational principles and the self-understanding of the traditional German hierarchical system of craftsmanship. This was dominated by highly skilled master craftsmen [*Meister*], assisted by experienced journeymen who guided and monitored apprentices and workers. The system was based on high quality of work considerations rather than adapting to the products' life span at the frontlines of just a few hours. Furthermore, the production process was hampered by huge fitting problems of parts delivered by multi-tiered supplier networks and, finally, by chaotic management from government and party organizations resulting in infighting and rapidly shifting or even contradictory decisions made by its top leaders (for solid documentation collated by domestic and international observers, see Overy 1980; Budraß 1998; Allen 2002; Uziel 2012). Despite these inherent problems, efforts to rationalize production in the aerospace industry had picked up by late 1941 and early 1942, with FWH Henschel in its Berlin-Schönefeld plant at the forefront until the end of the war (see Braun 1990 regarding specific technical challenges and Budraß 1998 for a detailed analysis of the years 1942 to 1944; for details with regard to the special strategies at FWH Henschel cf. Budraß & Grieger 1993).

Three critical developments forced decision makers to improve the efficiency of air armament production as of 1943: First, extreme labor shortages in the industry as even highly skilled specialists were increasingly drafted to compensate for the huge losses of troops at the eastern front (Moscow in late 1941, Stalingrad late 1942/early 1943); second, a rising vulnerability of the arms industry to Allied air raids and, third, the rapid increase of air combat attrition rates.

As these pressures soared toward the end of the war, a decisive step was to systematically dislocate production to unobtrusive surface sites or to areas still out of reach of the Allied Air Forces.

Another decisive step soon followed: Peenemünde, the site for the development and initial production of the A4 ballistic rocket, was partly destroyed in August 1943 by an RAF bomber command. Subsequently, the production was mainly relocated to a bombsafe underground mine called Mittelwerk at Kohnstein Hill close to Nordhausen on the southern slope of the Harz Mountains (State of

Thuringia, for details, see Section 4.4.1). It was soon propagated as a model project (Wagner 2004a).

Because of a now virtually complete Allied air supremacy, Germany's political and military leadership decided to relocate before long most of the aerospace industry to bombproof underground sites in 1944 and 1945. A case in point is the implementation of *Jägerstab*, the ad hoc task force charged with increasing fighter production, with a plan to construct about 170 underground relocations, often in mines or tunnels (yet mainly remaining incomplete before the end of the war). Final desperate steps included makeshift aircraft assembly sites under forest cover with camouflage nets, euphemistically called 'forest facilities' (e.g., Messerschmitt's *Waldwerke*, see Schmoll 1998, Annex 9, 10), and so-called 'bunker factories', that is, surface factories protected by enormous concrete vaulting (Raim 1992; Hartmann 2017b). Most of these projects required enormous and challenging construction work, carried out by forced labor from prisoner and concentration camps.

Thus, a number of originally independent evolutionary components, such as war developments, industrial rationalization processes and the ruthless exploitation of forced labor eventually, combined to form an interdependent and dehumanizing synergy that reached a dreadful climax during the last months of the war. Rationalization meant that a higher proportion of unskilled workers could be used (not initially the reason for employing forced labor).

4.3.1.2 *The commodification of misery in the late-war archipelago of subcamps*

So far, the concept of 'forced labor' has been used as a generic term to denote a part of the crude reality in Nazi Germany. Compulsory work was imposed on dissidents, members of other political parties, ethnic or religious minorities, illegally rounded-up civilians in the occupied countries, criminals or prisoners of war. Due to the catastrophic shortage of labor toward the end of the war, a steadily increasing number of concentration camp inmates were also forced to work for and eventually in the armament industries themselves.

A number of crucial political decisions in early 1942 marked the transition to a final radical phase of the regime's systematic utilization of forced labor. This included a pervasive reorganization of the armament industry, distinct management practices for the overall workforce in Germany and, last but not least, the authorization given to the SS organization in charge of its Business Administration [*Wirtschaftsverwaltunghauptamt/WVHA*] to proceed with a systematic commodification of camp inmates for the total armaments sector (for details, see Allen 2002). It must be emphasized that these decisions were also problematic from a National Socialist ideological point of view as they revealed the regime's ideological late-war dilemma: To save the weapons production, the SS and other National Socialistic organizations had to rely on precisely those people as an indispensable workforce who were supposed to be murdered in the gas chambers at the death camps but now were relentlessly killed through a policy of 'extermination by work' [*Vernichtung durch Arbeit*] mainly in the subcamps (for literature regarding these contradictions and complications, see Wagner 2004a:51 subseq.; Allen 2002:6 subseq.; Riexinger & Ernst 2003). Yet, the destinies of these groups depended on highly differentiated, often erratic rules of treatment as well as on widely varying strategies adopted by corporations or institutions rather than always being imposed by the SS security staff (see Buggeln 2012 for a well-documented and nuanced discussion).

Some industrial corporations had established production sites close to the main camps since the early 1940s. Others had their detainees work on parts or components in the camps proper (e.g., Messerschmitt at Flossenbürg) or in newly established camps close to a factory. In this respect, both Heinkel at Oranienburg and BMW at München-Allach soon became models for the subsequent evolution of this growing mutual interdependence of industrial production and forced labor. They got their forced labor from nearby main camps Sachsenhausen and Dachau, respectively. Toward the end of the war, however, countless subsidiary camps and 'commandos' [*Außenkommando*] were established, administered by the main camps. They were located close to the factories or even in sensitive production contexts, testifying to the fact underlined earlier that the industry's level of rationalization now increasingly allowed even inexperienced forced labor to work in what often must be regarded as high-tech facilities.

It is generally assumed that there were about 1,000 subcamps toward the end of the war. Germany and some of the occupied eastern countries were completely dotted with these sites of incarceration of forced labor. In late 1944, an estimated two-thirds of the 600,000 concentration camp inmates lived in subcamps. The camps were found not only in and near the arms factories but wherever labor was lacking. The number of concentration camp inmates working in the aerospace industry is estimated at 118,000 (Buggeln 2012:128). With some notable exceptions, both work and living conditions varied from outrageous to barely acceptable, depending on the personalities involved in decision-making and control (for a broad body of literature documenting these enormous differences, see Werner 2006, introductory chapter).

The preceding sections' description of the late war's entangled patterns of corporate *rationalization*, on the one hand (including spatial dispersal), and NS state terrorism with the *commodification of misery*, on the other, is certainly incomplete. But it is obvious that the far-reaching functional and spatial implications of this late-war evolution must be considered when the air armament's legacy in Germany (and beyond) is addressed from an industrial heritage perspective. So, the remaining Chapter 4 will discuss if and how the specific aspects of the NS industrial past are represented in transnational, national, community, corporate, civil society and even individual approaches.

4.4 Germany's landscapes of WWII aerospace commemoration: scales, sites and actors

4.4.1 Peenemünde/Mittelbau Dora as sites with national and transnational reach

In order to better understand stunning contrasts in Germany's contemporary landscapes of commemoration it is appropriate to first focus on the nationally and transnationally most prominent WWII sites for long-distance weapons development, yet together with its functionally connected late-war follower of mass production, the underground facility Mittelwerk at Nordhausen.

From a national perspective, the war's original test and early production site of the long-distance weapons Fi 103 and A4 at Peenemünde was known, mostly by hearsay, as one of the sites for so-called 'wonder weapons'. They would contribute, was the hope, to turning around the tides of the war. After the war, however, Peenemünde's reputation changed radically: It was now framed in Germany and beyond as 'the cradle of space travel' (see later), a narrative that was propagated successfully not least by the facility's most influential actors, among them Wernher von Braun and Walter Dornberger. The result was a pervasive post-war perception, not only in Germany, that Peenemünde could be characterized by the contrast of innocent scientists and engineers, on the one hand, and an evil politico-military leadership, on the other (Erichsen & Hoppe 2004; Hoppe 2004; Jikeli & Werner 2014; Eisfeld 2014).

This view purposefully withheld the now well-documented fact that the 'rocket people' at Peenemünde not only had used forced labor from the beginning of the war. Their leaders had even taken the initiative, as of 1943, to request the SS to send concentration camp detainees. Thus, thousands of forced laborers were exploited at Peenemünde, too, located in camps at Wolgast (just outside the island of Usedom) and, inside the strictly military compound, Karlshagen I and II, altogether an estimated 10,000 to 12,000 people (cf. Dornberger 1941; Erichsen & Hoppe 2004:339; Jikeli & Werner 2014).

Yet, Mittelwerk, with its roughly 40 sub-camps of Mittelbau-Dora, reflects even more strongly the Nazi regime's radicalization. There, an estimated 60,000 forced laborers from all over Europe had been abused from 1943 to April 1945, and more than 20,000 of them had died, by pure exhaustion, starvation or willful murder by SS guards (for a thorough documentation and appraisal, see Wagner 2004a, 2004b, 2009; Schafft & Zeidler 2011).

As briefly described earlier, Peenemünde and Mittelwerk/Mittelbau-Dora have been closely linked functionally. Thus, they should be understood together, a desideratum, however, that is difficult to achieve as they are not only located far away from each other but also administered by two different institutions in two different German states.

Today, each represents in its own way a dedicated, publicly funded and professionally managed site of remembrance. Both address in poignant ways the particularly dreadful reality of late-war industrial transnationalities, both with regard to inner functions and transnational reach (see the insightful assessments presented by Erichsen & Hoppe 2004; Allen 2002 and https://museum-peenemuende.de/; http://www.buchenwald.de/en/29/).[7] Regarding the link between German armament industries and forced labor, both sites of remembrance can be, arguably, considered the most important sites in Europe.

From a decidedly geographic industrial (heritage) perspective, however, a significant issue is still open: the inner functionalities and spatialities of the A4 production system and its supply-networks. Still dependent on research and development at Peenemünde, Mittelwerk functioned mainly as a 'site of assembly'. It was supplied by roughly 450 contract companies, relying on a conspicuous multitude of sub-contractors, delivering tens of thousands different single parts from Germany and occupied regions (Wagner 2004a:201 subseq.). This part of the late-war reality (and its enormous opportunity costs) remains a missing link for a deeper understanding, not only from a geographical perspective but also for many other disciplines and industrial heritage objectives alike.

Yet, the public fascination with Peenemünde and its A4 rocket is quite persistent – as it seems to be all but ubiquitous (see, for instance, Neufeld 1995; Petersen 2011; Barber 2017 with their remarks and references related to the German rocket specialists' destinies from Huntsville to Cape Canaveral). It represents a disturbing case in point for conflicted framings between awe and revulsion.

Nye (1994) has insightfully analyzed such disparate feelings in the face of technical feats with the concept of 'the technological sublime', also briefly referring to the A4 ballistic rocket. A partly overlapping, yet more general conceptual approach is that of 'technical emotions' (for a recent overview cf. the anthology edited by Heßler 2020). Seifert-Hartz (2020) presents illustrative findings related to technical emotions, based on thousands of entries in visitor books at Peenemünde, made by tourists who just had toured the site and its relics. Obviously, a whole range of conflicting emotions plague many visitors until today, such as fascination and abhorrence, fury and culpability.

By contrast, there is very little evidence that people working at Peenemünde or in its linked facilities were torn by comparable tensions, be it on the work floor or in the leadership, be it during the war or afterwards (for an exception, see Wegener 1996:95/135 subseq.; the memoir of a Peenemünde wind channel specialist). But how was it possible for them to become accomplices deeply involved in the Nazi regime's crimes, even if only a few of them fervently shared its ideology?

Neufeld (1995) was one of the first historians to painstakingly track the stunning pre-war, war and post-war itineraries of German rocket scientists as well as their technical progress and its political and military environments. Yet, and relying on increasingly accessible original documents in both the United States and Germany, he concluded, not at all surprising at the time, that "slave labor was the one uniquely 'Nazi' aspect of the rocket program" (op. cit. p. 277). In this way, he basically reproduced the narrative propagated by the aforementioned perpetrators, that is, the innocent rocket people, on the one hand, and the evil regime, on the other.

Only a few years later, however, a radical shift of questioning has led to new answers with regard to still-existing conundrums of the Nazi period, namely not asking, "Why were those involved not repulsed by their actions" but instead: "Why did they believe it was the right thing to do?" (Allen 2002:5, focusing on the mid-level leaders of the SS WVHA).

With regard to the rocket specialists, the issue maybe best epitomized by an angry Nazi insider's stunning comment, referred to in another renowned historian's treatise: "At Peenemünde, They Have Created a Paradise" (Petersen 2011:64). This author, as Allen did, delved into the 'history of everyday life' (in German: *Alltagsgeschichte*), looking into ways of enculturation in a closely knitted specialist community with an extreme devotion to and identification with their work (op.cit. p. 3). Peenemünde was a case in point, allowing to live like privileged members of a secluded club: Confined to a place as isolated as idyllic, marked by secrecy, coercion and consent as well as by mutual trust, understanding and reliability (op.cit. p. 81). Furthermore, it was far from the risks and horrors of dire frontlines and, last, but not least, the employees were spoiled by high salaries and a growing feeling of self-worth. In his conclusion, Petersen stated (op.cit. p. 249): "Half military facility, half technological Shangri-la, Peenemünde created a cultural environment in which the needs of the regime and the needs of the missile specialists were inseparably intertwined".

This background information about Peenemünde is not only historically highly intriguing, it will also be crucial for interpretational strategies in the context of future industrial heritage issues in this field.

If even Peenemünde and Mittelwerk/Mittelbau-Dora cannot satisfy all aspirations of a geographically-inspired industrial heritage gaze, one major question remains: Then how is the late-war aerospace industry represented in Germany's remaining landscapes of memory generally?

4.4.2 National memorial sites of Nazi crime victims in Germany

To date, and setting aside historical treatises and civil society commemoration work, there has not been any official and systematic stocktaking of the late-war aerospace sector's heritage sites in Germany. Two extensive volumes published by Federal Agency for Civic Education (*Bundeszentrale für Politische Bildung* 1995, 1999) in the national Ministry of the Interior produced a thorough documentation of Nazi crime-related heritage sites of the States of West Germany (vol. 1) and of the States of former East German (vol. 2), respectively. A careful examination leads to a surprising insight: Only a small percentage of thousands of entries refer to the armament industry generally, and only a few dozen do so with regard to individual production facilities or companies of war aircraft or aircraft engines. A thematic map of the memorial sites for both parts of re-united Germany (*Gedenkstätten für die Opfer des Nationalsozialismus in der Bundesrepublik Deutschland* published in 1999) chose to showcase eight types of memorial sites. One category, the places associated with the subsidiary labor camps and commandos (where armament production in the late WWII years occurred), is incomplete, though.

The lack of inclusion of hundreds of places was discussed by several German historians in the 2000s. Schalm (2009) argues it may have been the long-time disregard of oral history, specifically of autobiographical accounts by the prisoners held there, that was the main reason for the neglect. In her work on the extensive Dachau subsidiary camp system Schalm points to the insufficient documentation of external camps in the reconstruction of the Holocaust (Schalm 2009:11–24). German historians working in collaboration with the United States Holocaust Memorial Museum on the *Encyclopedia of Camps and Ghettos, 1933–1945* (Megargee 2009) contributed to a more comprehensive list of camps associated with the armament industry in Germany. Though, the aerospace industry with its excessive late-war use of forced labor under SS administration remains less visible.

4.4.3 Community patterns of commemoration

A small selection of civic communities with locations at the larger aerospace corporation sites may serve as exemplary cases, that is, Dessau (Junkers), Rostock and Oranienburg (both Heinkel), Berlin and Kassel (both Henschel) as well as Augsburg and Regensburg (both Messerschmitt).

Despite considerable differences in size, all these firms mentioned implemented large-scale processes of industrialization at their main locations from the early to the late 1930s, accompanied by major infrastructural developments. Some of these still exist today, despite the heavy war destruction and post-war dismantling of the industrial facilities by the victorious powers. At almost all locations, there are substantial tangible legacies. Notable sites, well preserved or repaired, include the former Messerschmitt apprentice workshop in Regensburg, the Henschel company's former main administration building in Kassel and office buildings at Berlin-Schönefeld airport.

Visually most striking, however, are the lasting traces of transportation networks (construction/extension of existing airports, streets, rail- and tramways, bridges) and of an enormous expansion of company housing. Much of it can be found in residential districts inhabited to this day, such as the Thomas-Müntzer-Platz neighborhood in Rostock as well as Weiße Stadt and Leegebruch districts in Oranienburg (both Heinkel) or the Ganghofer monument complex for the Messerschmitt workforce in Regensburg-Prüfening.

Furthermore, both tangible and intangible aspects are documented in specific museums, exhibition halls and archives, focusing specifically on the companies themselves. But as the latter are no longer existing as independent entities (with the exception of Henschel, see later), sites still featuring original names of former aircraft producers are established by private initiatives, such as the foundations of Dornier or Messerschmitt, the latter in cooperation with Airbus (see later), such as:

- **Junkers:** Junkers Technikmuseum 'Hugo Junkers', Dessau, today a subdivision of the town of Dessau-Roßlau, State of Saxony-Anhalt (https://technikmuseum-dessau.org/)
- **Messerschmitt:** Flugmuseum Messerschmitt, Manching, State of Bavaria (https://www.flugmuseum-messerschmitt.com/)
- **Heinkel:** Museum Remshalden, State of Baden-Württemberg (https://www.museumsvereinremshaldenvereinde/heinkel)
- **Dornier:** Dornier Museum Friedrichshafen, State of Baden-Württemberg (https://www.dorniermuseum.de), and, remembering the former Air Force's most important site research and test centers
- **Luftfahrttechnisches Museum Rechlin e.V.**, State of Mecklenburg-Vorpommern (https://www.luftfahrttechnisches-museum-rechlin.de/)

The Henschel museum in Kassel is not listed here as former military production is not presented or shown in more detail. A number of successor companies have emerged from former divisions of Henschel, now mostly integrated into other enterprises

Furthermore, a number of original planes in particular from WWII are exhibited at some of the places mentioned earlier, both in Germany and abroad (e.g., Deutsches Museum, München; Deutsches Technikmuseum Berlin; Militärhistorisches Museum, airport Berlin-Gatow as well as Royal Airforce Museum, Cosford, UK; National Air and Space Museum, USA).

Even a cursory analysis will show that the places in Germany listed earlier mirror the same deficiencies in the commemoration of the events: Even in cities and former industrial clusters that constituted the nodes of the German aerospace industry hardly any information exists that would allow an understanding of late WWII production systems, let alone the intricate organizational and spatial patterns as well the mutually constitutive links between production and forced labor (beyond the memorial sites listed in the documentations analyzed earlier).

If widely known tangible or intangible aspects are named at all (neighborhood designations, buildings, ruins, street names), their dark entangled state/community/corporate implications are all but ignored, and consciously disregarded. Obviously, more research is desirable, but as it seems now most places in Germany with a strong formative past of the aerospace industry mirror highly inadequate commemoration patterns.

4.4.4 Corporate patterns of commemoration

Current strategies of the aerospace and automotive companies that were formerly enmeshed in rationalization/forced labor use reflect the deficiencies in the commemoration practices described earlier. Yet, there is a big difference when it comes to the visibility of the different industrial sectors: Some of the main corporations of the aerospace industry such as Junkers, Messerschmitt, Heinkel or Dornier continued to exist. However, company productions had shrunk considerably. As the building of aircraft was interdicted immediately after the war the former aircraft corporations specialized during the early post-war period in the production of a large variety of civil goods, such as motorcycles, small cars or household equipment.

By contrast, the main automotive corporations survived and simply continued under their original names (sometimes slightly changed) almost immediately after the war, not least in order to satisfy the Allied forces' needs (repair, renewal of motor vehicles fleet). Within a few years, they experienced a spectacular growth of passenger cars, too.

Not until post-war constraints were lifted in the context of West Germany's accession to NATO in May 1955, some of the main former aerospace corporations reorganized to catch up: They quickly became internationally competitive again in highly-specialized military and civil aerospace production sectors, such as helicopter development or satellite construction. Subsequently, however, the firms became all but invisible (except for the historic designations mentioned earlier): Successive consolidations and mergers took place, increasingly not only with other German but also with European partners, in particular France. Eventually, this process resulted in aerospace groups such as DASA (Deutsche Aerospace AG) and

EADS (European Aeronautic Defence and Space Company NV), the predecessors of today's Airbus. It is understandable that the latter transnational company is not really interested in commemorating its German precursors at their former sites – yet with a remarkable exception: The already mentioned Flugmuseum Messerschmitt in Manching.

By contrast, the originally constitutive automotive sector of WWII's aerospace industry, most prominently Audi, BMW, Daimler and Volkswagen, cannot hide its history in the same way. Instead, these firms have adopted a different strategy to deal with their past. For decades, they have chosen to let aspects of their uncomfortable past fade away. . .. Not least, this was achieved by creating highly innovative corporate brand worlds at their headquarters, especially by Volkswagen and BMW, with VW Autostadt in Wolfsburg and BMW World in München, respectively (cf. https://www.bmw-welt.com/en.html and https://www.autostadt.de/en/start; for a critical appreciation of Autostadt, see Soyez 2016).

At Volkswagen, the commemoration of the difficult past is relegated to one of its traditional factory buildings in Wolfsburg. It is located in its basement with ongoing production above. Underneath a small museum has been established to honor the memory and sufferings of the facility's forced labor workforce. Here, a few personal items of former forced laborers as well as pieces of the weaponry produced there are presented (see Volkswagen AG/Historische Kommunikation 1999). As this place of commemoration in the center of ongoing production is quite difficult to access, a relocation of this museum to the corporation's Autostadt brand world would be appropriate, thus making it potentially accessible to millions of visitors every year. Yet, it seems that such a move has never been seriously considered.

BMW, also heavily entangled in WWII's aerospace industry, has also adopted a split strategy: While its brand world named BMW World in München, inaugurated in 2007, hardly addresses war involvements, a dedicated Place of Remembrance was opened in the adjacent Museum in 2020, clearly documenting its WWII armament production. The focus is on its former aircraft engine factory at München-Allach with a large camp housing thousands of forced laborers (https://www.bmwgroup.com/en/company/history/BMW-during-the-era-of-national-socialism.html;   cf. also Frankonzept 2018, a copiously documented expertise).

In this context, the corporations of Audi and Daimler must be regarded as laggards. Their brand worlds Audi Forum (Ingolstadt) and Daimler Museum (Untertürkheim, close to Stuttgart) show no consistent documentation of their WWII involvements.

4.4.5 Civil society approaches to commemoration

The Society for the Preservation of Historic Sites of German Aviation (*GBSL: Gesellschaft zur Bewahrung von Stätten deutscher Luftfahrtgeschichte, www.luftfahrtstaetten.de*) in Berlin deserves to be mentioned here.

The non-profit society's achievements include a most impressive national inventory of historic sites of the German aerospace industry, as a whole and since its beginnings. The inventory has been compiled since 1992; the results have been published in the form of two folded maps showing more than 3,000 sites nationwide. This inventory is mainly based on questionnaires returned by hundreds of associations, institutions and individuals working with the large community of German Aviation History specialists and enthusiasts. While the entry forms and contents widely differ and comprise sites from different eras (they also include historic sites of the German aviation during the WWII years), three types of information are common:

- *Actor groups*: It is shown from whom the varied data originated, whether it derived from authorities, institutions, associations, civic groups or individuals
- *Heritage status*: Whether the sites gained recognition on a national or local level
- *Conclusions* about the main character and relevance of the marked sites, an ongoing form of pioneering stock-taking of German aviation heritage.

Two more examples of civic engagement mapping strong transnational links from the national to the local levels need to be highlighted. The first one was the result of a study limited to a district: A list of sites in the eastern part of the State of North Rhine Westphalia which was compiled with the help of the aforementioned GBSL society. The focus of regional events in Germany's aviation history may seem to be insignificant. However, it shows the breadth of commemoration in places (so-called *Erinnerungsmale*); the heritage sites – often managed by private actors – are remembered in the form of small plaques with inscriptions, stones, steles or sculptures (Hoebel 2019, https://www2.lwl.org/de/).

The second example represents a kind of a micro case study of a WWII subsidiary camp in Saalfeld in Thuringia (with the codename Laura administered by the Buchenwald Concentration Camp). Here, A4 rocket power units were tested in underground facilities, mainly relying on the work of over 2,000 forced laborers from more than ten European countries. The Laura camp was operating from late 1943 to the end of the war. Early studies of the sites in the 1960s were initiated by the Young Historians, a group of high school students. They helped to produce a series of newspaper articles in the local press and created the first commemorative marker. More formal studies were taken up again by a Friends' Association (Förderverein Gedenkstätte Laura e.V.) which was formed in 1998. The first comprehensive history of the site was published in 1999 (Gropp); the research efforts with increasing local support and publications by former inmates and professional historians resulted in a new memorial center with revised exhibits. Finally in 2012, a publicly funded and staffed memorial museum was opened. The commemorative studies at Laura are amazing example of committed, transnational remembrance work (see Schmidt van der Zanden 2009 and website of the Laura Memorial https://www.kz-gedenkstaette-laura.de/en/about-us/).

4.5 Conclusion

In less than a generation, a major shift in influential public discourse has led to a widespread interest in the preservation of industrial heritage in Germany. Within only a few years an impressive number of sites and even a few contiguous landscapes of commemoration are now under protection in various categories.

Yet, delicate shortfalls persist in particular with regard to industrial development during dark historical contexts, time-space processes in multi-tiered and multi-local supply networks as well as ensuing domestic and transnational patterns of interaction, all of this documented in widespread functionalities and spatialities.

The heritage-related inventory of Germany's late WWII aerospace industry, to date a largely overlooked lead sector, was conducted in two steps. First, the focus was put on two originally independent strands of its time that started to merge in 1942 already, that is, the technical rationalization of the industry, on the one hand, and forced labor, on the other. Second, their representation in contemporary landscapes of commemoration was examined.

The results of this investigation are as clear-cut as they are disturbing: Except for a few remarkable sites the national and transnational functionalities and spatialities of Germany's late-war aerospace industry are absent from the country's otherwise rich landscapes of remembrance. Obviously, this memory is still too delicate.

Consequently, the uncomfortable, systemic and transnational aspects in former WWII settings deserve a much more focused attention and recognition.

Acknowledgment

I am very grateful to Rudi Hartmann for critical remarks on an earlier draft and valuable suggestions, as well as to Christine O'Neill for improving my English.

Notes

1 The main focus will be on the German aerial armament industry and its links with forced labor (as to the term, see next note. The beginnings and development of the space industry, mainly linked to the names of Wernher von Braun and the Army's Research Center at Peenemünde with the development of a ballistic rocket A4 (abbreviation of the originally used term Aggregat 4) are also characterized by the (highly secretive) inclusion of crucial plane and car producers, here in particular their air engine subsidiaries (for details regarding early links with the industry, cf. Petersen 2011:57/58).

2 Although systematically used in authoritative American treatises (see, for instance, Neufeld 1995, Allen 2002, Petersen 2011), the term 'slave labor' is discarded here intentionally and replaced by the more generic term 'forced labor'. The treatment of camp detainees could vary enormously, even in the infamous late-war subcamps (cf. Buggeln 2012 for an update of the academic discussion with many references to the aerospace industry). Other terms often used in the international literature are also discarded in the following, that is, the Nazi jargon for so-called 'vengeance weapons' (V1, V2). Their original technical designations were instead Fi 103 (named after its main developer firm Fieseler) and A4.

3 While most of these publications represent what could be called academic discourses, another facet of today's reality should not be ignored: The impressive contributions of civil-society associations and individual actions, normally not clearly visible in the references mentioned above. This multi-facetted world of (mostly) bottom-up initiatives, often presented in what is called 'grey literature', can only be presented very briefly at the end of the chosen case studies (Section 4.4.5).

4 Messerschmitt aircraft factory, originally a branch plant of Bayerische Flugzeug-
   werke (BFW) in Augsburg, was constructed 1936/1937 in the western part of the city
   of Regensburg, local district of Prüfening (State of Bavaria), together with more than
   1,000 residences for the workforce. Soon, the facility had become one of the largest and
   most performant aircraft factories in Germany, mainly producing the fighters of the Me
   109 series, altogether some 11,000 planes before the end of the war, mostly with the
   help of thousands of forced labor (cf. Schmoll 1998). While the aircraft factory was all
   but destroyed after 1943, large parts of the company housing were not or only slightly
   affected and are today preserved under monument protection. Yet, the darker aspects were
   only addressed recently, see Note 6).
5 The author is very grateful to the Board of the GBSL Society for allowing him to sift
   through the original files of this inventory. As exemplary states were selected: The States
   of Bavaria, Brandenburg, Mecklenburg-Vorpommern, Saxony-Anhalt and Thuringia, a
   total of almost 1,000 entries. These latter were very diverse but allow the conclusions
   presented in Section 4.5 (cf. Schmitt 1992).
6 The Regensburg case, mentioned earlier, can also be presented in this chapter on civil
   courage: This city represents a most illustrative case in point for the systematic suppres-
   sion of its late war memories by city officials, both administration and heritage authorities as well as political parties. This long-lasting silence was eventually broken by civil
   society actors, leading to a slow rethinking of heritage approaches to Nazi and WWII (see
   https://www.regensburg-digital.de/messerschmitt-und-regensburg/13052012/).
7 As of 1938, the Peenemünde area was split up into two separate, yet cooperating, administrative entities: The larger state-owned Army research area/HVA, mainly focused
   on the development of the ballistic rocket A4 and the Air Force Research Center of
   Peenemünde-West, occupying the much smaller northwestern part of the Usedom
   island. Both areas are today widely protected heritage landscapes which are open to
   visits and guided tours (see overview map: https://museum-peenemuende.de/zeitreise/
   denkmallandschaft/).

References

Allen, M.T. (2002) *The business of genocide: The SS, slave labor, and the concentration
    camps.* Chapel Hill and London: The University of North Carolina Press.
Ashworth, G.J. & Hartmann, R. (Eds.). (2005) *Horror and human tragedy revisited: The
    management of sites of atrocities for tourism.* New York: Cognizant.
Barber, M.R. (2017) *V2: The A4 rocket from Peenemunde to Redstone: Design – develop-
    ment – operations.* London: Ian Allan Ltd.
Beauvais, H., Kössler, K., Meyer, M. & Regel, C. (1998) *Flugerprobungsstellen bis
    1945 – Johannisthal, Lipezk, Rechlin, Travemünde, Tarnewitz, Peenemünde-West.* Bonn:
    Bernard & Graefe (= Die deutsche Luftfahrt – Entwicklungsgeschichte der deutschen
    Luftfahrttechnik).
Benz, W. & Distel, B. (Eds.). (2006) *Der Ort des Terrors: Geschichte der nationalsozi-
    alistischen Konzentrationslager.* München: C.H. Beck (= Vol. 1, Die Organisation des
    Terrors).
Braun, H.-J. (1990) Fertigungsprozesse im deutschen Flugzeugbau 1926–1945. *Technikge-
    schichte, 57,* 111–136.
Budraß, L. (1998) *Flugzeugindustrie und Luftrüstung in Deutschland 1918–1945.* Düssel-
    dorf: Droste Verlag (= Schriften des Bundesarchivs 50).
Budraß, L. (2016a) *Adler und Kranich Die Lufthansa und ihre Geschichte 1926–1955.*
    München: Blessing.
Budraß, L. (2016b) *100 Jahre BMW: Die Schatten der NS-Vergangenheit.* https://www.
    dw.com/de/100-jahre-bmw-die-schatten-der-ns-vergangenheit/a-19094556.

Budraß, L. & Grieger, M. (1993) Die Moral der Effizienz. Die Beschäftigung von KZ-Häftlingen am Beispiel des Volkswagenwerks und der Henschel Flugzeug-Werke. *Jahrbuch für Wirtschaftsgeschichte/Economic History Yearbook, 34,* 89–136.

Buggeln, M. (2012) *Das System der KZ-Außenlager Krieg, Sklavenarbeit und Massengewalt.* Bonn: Friedrich-Ebert-Stiftung (= Reihe Gesprächskreis Geschichte, Heft 95).

Bundeszentrale für Politische Bildung. (1995) *Gedenkstätten für die Opfer des Nationalsozioalismus: Eine Dokumentation. Band I: Baden-Württemberg, Bayern, Bremen, Hamburg, Hessen, Niedersachsen, Nordrhein-Westfalen, Rheinland-Pfalz, Saarland, Schleswig-Holstein* (Autoren: Martin Stankowski/Ulrike Puvogel [Redaktion], unter Mitarbeit von Ursula Graf), Bonn, Dez. 1995. (Basierend auf: Bundeszentrale für Politische Bildung, Bonn 1987 [Schriftenreihe der Bundeszentrale Band 245]).

Bundeszentrale für Politische Bildung. (1999) *Gedenkstätten für die Opfer des Nationalsozioalismus: Eine Dokumentation Band II: Bundesländer Berlin, Brandenburg, Mecklenburg-Vorpommern, Sachsen-Anhalt, Sachsen, Thüringen* (Autoren: Ulrike Puvogel [Redaktion] Stefanie Endlich, Regina Scheer, Beatrix Herlemann, Nora Goldenbogen, Monika Kahl), mit Thematischer Karte, Bonn 1999.

Coe, N.M. & Yeung, H.W.-C. (2015) *Global production networks: Theorizing economic development in an interconnected world.* Oxford: Oxford University Press.

Dicken, P. (1992) *Global shift: The internationalization of economic activity.* London: Chapman Publishing (1st edition).

Dicken, P. (2015) *Global shift: Mapping the changing contours of the world economy.* Los Angeles: Sage (7th edition).

Dornberger, W. (1941) Betr. Entwicklungsgrundsätze (Schreiben an Heeresversuchsanstalt Peenemünde, Berlin, 13. November 1941). In Erichsen, J. & Hoppe, B.M. (Eds.), *Peenemünde: Mythos und Geschichte der Rakete 1923–1989*, pp. 346–347. Berlin: Nicolai (= Katalog des Museums Peenemünde).

Eisfeld, R. (2014) Der "Mythos Peenemünde". In Jikeli, G. & Werner, F. (Eds.), *Raketen und Zwangsarbeit in Peenemünde*, pp. 218–250. Schwerin: Friedrich-Ebert-Stiftung.

Erichsen, J. & Hoppe, B.M. (Eds.). (2004) *Peenemünde: Mythos und Geschichte der Rakete 1923–1989.* Berlin: Nicolai (= Katalog des Museums Peenemünde).

Frankonzept. (Ed.). (2018) *Machbarkeitsstudie Dokumentationsstrategie KZ-Außenlager Allach.* Würzburg: Frankonzept.

Graham, B., Ashworth, G.J. & Tunbridge, J. E. (2000) *A geography of heritage – power, culture and economy.* London: Arnold.

Gregor, N. (1998) *Daimler-Benz in the Third Reich.* New Haven and London: Yale University Press.

Gropp, D. (1999) *Aussenkommando Laura und Vorwerk Mitte Lehesten – Testbetrieb für V2-Triebwerke.* Berlin and Bonn: Westkreuz-Verlag.

Hamburger Stiftung für Sozialgeschichte des 20. Jahrhunderts (Ed.) (1987) *Das Daimler-Benz-Buch. Ein Automobilkonzern im "Tausendjährigen Reich".* Nördlingen: F. Greno.

Hartmann, R. (2013) Dark tourism, thanatourism, and dissonance in heritage tourism management: New directions in contemporary tourism research. *Journal of Heritage Tourism, 9,* 166–182.

Hartmann, R. (2017a) Changing memorial landscapes, changing approaches in the study of memorial sites for the victims of national socialist Germany: A review. *Berichte zur deutschen Landeskunde, 91,* 349–372.

Hartmann, R. (2017b) Places with a disconcerting past: Issues and trends in Holocaust tourism. *Europe Now/Council for European Studies (CES).* www.europenowjournal.org/2017/09/05/places-with-a-disconcerting-past-issues-and-trends-in-holocaust-tourism/.

Hayter, R. & Patchell, J. (2016) *Economic geography: An institutional approach.* Don Mills, ON: Oxford University Press Canada (2nd edition).

Heßler, M. (Ed.). (2020) *Technikemotionen.* Paderborn: Ferdinand Schöningh.

Hirschel, E.H., Prem, H. & Madelung, G. (2004) *Aeronautical research in Germany – from Lilienthal until today*. Berlin: Springer.

Hoebel, C. (2019) *Topographie des Fliegens/Westfalen Lippe: Erinnerungsmale – Inventar*. Münster: Thiekötter Druck.

Holland, J. (2018) *Big week: The biggest air battle of World War II*. New York: Atlantic Monthly Press.

Hopmann, B., Spoerer, M., Weitz, B. & Brüninghaus, B. (1994) *Zwangsarbeit bei Daimler-Benz*. Stuttgart: Franz Steiner Verlag.

Hoppe, B.M. (2004) Peenemünde: Ein Beitrag zur deutschen Erinnerungskultur. In Erichsen, J. & Hoppe, B.M. (Eds.), *Peenemünde: Mythos und Geschichte der Rakete 1923–1989*, pp. 11–22. Berlin: Nicolai (= Katalog des Museums Peenemünde).

Jähner, H. (2021) *Aftermath: Life in the fallout of the Third Reich, 1945–1955*. New York: Penguin Random House (German edition published in 2019, under the title: Wolfszeit. Deutschland und die Deutschen 1945–1955. Berlin: Rowohlt).

Jeffreys, D. (2010) *Hell's cartel: IG Farben and the making of Hitler's war machine*. New York: Holt.

Jikeli, G. & Werner, F. (Eds.). (2014) *Raketen und Zwangsarbeit in Peenemünde*. Schwerin: Friedrich-Ebert-Stiftung.

Kukowski, M. & Boch, R. (2014) *Kriegswirtschaft und Arbeitseinsatz bei der Auto Union AG Chemnitz im Zweiten Weltkrieg*. Stuttgart: Franz Steiner Verlag (= Beiträge zur Unternehmensgeschichte 34).

Li, L. & Soyez, D. (2016) Transnationalizing industrial heritage valorizations in Germany and China – and addressing inherent dark sides. *Journal of Heritage Tourism*, *12*, 296–310.

Logan, W. & Reeves, K. (Eds.). (2009) *Places of pain and shame: Dealing with 'difficult heritage'*. Milton Park: Routledge.

Megargee, G.P. (Ed.). (2009) *Encyclopedia of camps and ghettos 1933–1945. Vol. 1, part A and part B: Early camps, youth camps and concentration camps and subcamps under the SS-business administration office (WVAH)*. Bloomington: Indiana University Press (= Published in Association with The United States Holocaust Memorial Museum.

Mommsen, H. & Grieger, M. (1996) *Das Volkswagenwerk und seine Arbeiter im Dritten Reich*. München: Econ.

Neufeld, M.J. (1995) *The rocket and the Reich: Peenemünde and the coming of the ballistic missile era*. Washington, DC: Smithsonian Books.

Nye, D.E. (1994) *American technological sublime*. Cambridge, MA: MIT Press.

Overy, R.J. (1980) *The air war 1939–1945*. London: Europa Publications Ltd. (Reprint: 2020).

Petersen, M.B. (2011) *Missiles for the fatherland: Peenemünde, national socialism, and the V-2 missile*. Cambridge, UK: Cambridge University Press (= Paperback edition, original issue in 2009).

Pohl, H., Habeth, S. & Brüninghaus, B. (1986) *Die Daimler-Benz AG in den Jahren 1933 bis 1945*. Wiesbaden: Steiner.

Raim, E. (1992) *Die Dachauer KZ-Außenkommandos Kaufering und Mühldorf: Rüstungsbauten und Zwangsarbeit im letzten Kriegsjahr 1944/45*. Landsberg am Lech: Landsberger Verlagsanstalt Martin Neumeyer.

Raim, E. (1998) Gescheiterte Gedenkinitiativen. Die Beispiele Kaufering und Landsberg. In Bannasch, B. & Hahn, H.J. (Eds.), *Darstellen, Vermitteln, Aneignen: Gegenwärtige Reflexionen des Holocaust (2018)*, pp. 415–432. Göttingen: Vienna University Press.

Riexinger, K. & Ernst, D. (2003) *Vernichtung durch Arbeit. Rüstung im Bergwerk. Die Geschichte des Konzentrationslagers Kochendorf – Außenkommando des KZ Natzweiler-Struthof*. Tübingen: Silberburg.

Schafft, G. & Zeidler, P. (2011) *Commemorating hell: The public memory of Mittelbau-Dora*. Urbana, Chicago and Springfield: University of Illinois Press.

Schalm, S. (2009) *Ueberleben durch Arbeit? Aussenkommandos und Aussenlager des KZ Dachau 1933–1945, Geschichte der Konzentrationslager 1933–1945, Band 10*. Berlin: Metropol.

Schmidt van der Zanden, C. (2009) Saalfeld ["Laura"]. In Megargee, G.P. (Ed.), *Encyclopedia of camps and ghettos 1933–1945*, Volume 1, Part A, pp. 411–413. Bloomington: Indiana University Press.

Schmitt, G. (1992) Aspekte der Bewahrung von Stätten deutscher Luftfahrgeschichte. In *Schriftenreihe zur Luftfahrgeschichte, Heft 1: "Traditionsbewahrung heute"*, pp. 1–9. Berlin: Gesellschaft zur Bewahrung von Stätten deutscher Luftfahrtgeschichte e.V./GBSL.

Schmoll, P. (1998) *Die Messerschmitt-Werke im Zweiten Weltkrieg: Die Flugzeugproduktion der Messerschmitt GmbH Regensburg von 1938 bis 1945*. Regensburg: Mittelbayrische Druck- & Verlagsgesellschaft.

Seifert-Hartz, C. (2020) "Vom Staunen über Entsetzen". Museumsbesucher und ambivalente Technikemotionen in Peenemünde. In Heßler, M. (Ed.), Technikemotionen, pp. 201–228. Paderborn: Ferdinand Schöningh.

Soyez, D. (2009) Europeanizing industrial heritage in Europe: Addressing its transboundary and dark sides. *Geographische Zeitschrift, 97*, 43–55.

Soyez, D. (2016) Industriekultur als städtisches Erbe und lebendige Präsenz: Selektions- und Interpretationsstrategien aus geographischer Sicht mit einem Ausblick auf Wolfsburg. In *Informationen zur Modernen Stadtgeschichte (Themenschwerpunkt Städtisches Erbe – Urban Heritage)*, pp. 55–67, January. Berlin: Deutsches Institut für Urbanistik.

Steinbacher, S. (2005) *Auschwitz: A history*. London: Penguin Books.

Steinecke, A. (2021) *Dark tourism: Reisen zu Orten des Leids, des Schreckens und des Todes*. München: UVK Verlag.

Tooze, A. (2006) *The wages of destruction: The making and breaking of the Nazi economy*. London: Allen Lane.

Uziel, D. (2012) *Arming the Luftwaffe: The German aviation industry in World War II*. Jefferson, NC and London, UK: McFarland & Company, Inc. Publishers.

Uziel, D. (2014) Zwangsarbeit am Fließband: Das "Produktionswunder" der deutschen Luftfahrtindustrie 1944. In von Lingen, K. & Gestwa, K. (Eds.), *Zwangsarbeit als Kriegsressource in Europa und Asien*, pp. 265–282. Paderborn: Ferdinand Schöningh.

Vajda, F.A. & Dancey, P. (1998) *German aircraft industry and production 1933–1944*. Warrendale, PA: SAE International.

Volkswagen AG/Historische Kommunikation. (1999) *Erinnerungsstätte an die Zwangsarbeit auf dem Gelände des Volkswagenwerks*. Wolfsburg: Volkswagen Aktiengesellschaft.

Wagner, J.-C. (2009) Mittelbau subcamp system In Megargee, G.P. (Ed.), *Encyclopedia of camps and ghettos 1933–1945*, Vol. 1, Part B, pp. 973–974. Bloomington: Indiana University Press.

Wagner, J.-C. (2004a) *Produktion des Todes. Das KZ Mittelbau-Dora*. Göttingen: Wallstein (2nd edition).

Wagner, J.-C. (2004b) Opfer des Raketenwahns. Zwangsarbeit in Peenemünde und Mittelbau-Dora. In Erichsen, J. & Hoppe, B.M. (Eds.), *Peenemünde: Mythos und Geschichte der Rakete 1923–1989*, pp. 43–52. Berlin: Nicolai (= Katalog des Museums Peenemünde).

Wegener, P.P. (1996) *The Peenemünde wind tunnels: A memoir*. New Haven and London: Yale University Press.

Werner, C. (2006) *Kriegswirtschaft und Zwangsarbeit bei BMW*. München: R. Oldenbourg Verlag (= Perspektiven, Schriftenreihe der BMW Group – Konzernarchiv, Vol. 1, im Auftrag von MTU Aero Engines und BMW Group).

5   Amsterdam under Nazi Germany Occupation Remembered (1940–1945)

Rudi Hartmann

On May 10, 1940, Nazi Germany invaded the Netherlands despite the country's neutral status. After the bombing of Rotterdam, with close to 1,000 civilian fatalities, the Dutch military forces surrendered on May 15, 1940. The Royal Family, including Queen Wilhelma, fled to England, where they formed a government in exile.

For five long years the residents of Amsterdam were subjected to the occupation of their city. A civilian administration dominated by the SS was put in place. As the Germans ruled initially with 'velvet gloves', resistance was at first carried out only by a minority. It grew in the following years. Towards the end of the occupation, when conditions deteriorated leading to starvation and the chronic lack of resources in a 'hunger winter' 1944/1945, resistance became widespread. In May 1945, Amsterdam and Holland were at last liberated by allied (Canadian) forces.

The turning point in the residents' attitude to the occupiers was the year 1941. On February 21, 1941, the Germans arrested several hundred Jews and deported them from Amsterdam first to the Buchenwald concentration camp and then to the Mauthausen concentration camp. Almost all of them were murdered in Mauthausen. The arrests and the brutal treatment shocked the population of Amsterdam. In response, Communist activists organized a general strike on February 25 and were joined by many other worker organizations. Major factories, the transportation system, and most public services came to a standstill. The Germans brutally suppressed the strike after three days, crippling Dutch resistance organizations in the process. In July 1942, the Germans began mass deportations of Jews to killing centers in occupied Poland, primarily to Auschwitz and to Sobibor. The city administration, the Dutch municipal police, and Dutch railway workers all cooperated in the deportations, as did the Dutch Nazi Party (NSB).[1] German and Dutch Nazi authorities arrested Jews in the streets of Amsterdam and took them to the assembly point for deportations – the Hollandsche Schouwburg, the theater building in the Jewish quarter.

Amsterdam, the country's largest city, had a Jewish population of about 75,000, which increased to over 79,000 in 1941 according to a survey of registered persons. Jews represented close to 10 percent of the city's total population. More than 10,000 of these were foreign Jews who had found refuge in Amsterdam in the 1930s. Among them were a large number of German Jews who, like the Frank

DOI: 10.4324/9780367823795-8

Family, migrated to Amsterdam. The German occupying powers confiscated the property left behind by deported Jews. In 1942 alone the contents of nearly 10,000 apartments in Amsterdam were expropriated by the Germans and shipped to Germany. Some 25,000 Jews, including at least 4,500 children, went into hiding to evade deportation. About one-third of those in hiding were discovered, arrested, and deported. In all, about 80 percent of the prewar Dutch Jewish community perished.[2]

What are the major preserved sites in the City of Amsterdam to commemorate the lives of the victims and the tragic events? Which are the leading museums that educate the public about the 1940–1945 period? The following museums and historic sites stand out:

- The Anne Frank House, with the secret annex where eight individuals were in hiding and young author Anne Frank wrote her diary 1942–1944
- The Dutch Resistance Museum (Verzetsmuseum)
- Four museums and/or historic sites managed by the Jewish Cultural Quarter (Joods Cultureel Kwartier): the Jewish Museum (Joods Museum), the Portuguese Synagogue, the National Holocaust Museum, and the Hollandsche Schouwburg

5.1 Anne Frank House (263 Prinsengracht)

The narrow house 263 Prinsengracht in the mixed-use Jordan neighborhood was the production site of two small businesses set up by Otto Frank. He had left his native Frankfurt, Germany, home for Amsterdam in 1933 when the Nazis seized power. He considered the Netherlands a safe place for his family and business ventures. The four-story house on the Prince Canal was divided into two parts, the front part with the offices, production site, and storage of the small business products; the back was hidden away from the front as a residential section. It was in this secret annex where eight individuals, all Jews originally from Germany, went into hiding in 1942; it was where Anne Frank wrote her diary. On August 4, 1944, the group in hiding was betrayed, arrested, and first deported to the Westerbork transit camp and from there on to Auschwitz on September 3.

Anne Frank's diary 'Het Achterhuis' (in Dutch) was published posthumously in 1947 by editor father Otto Frank, the only survivor of the group in hiding. The book was a moderate success and translated into German, French, and English in the following years. The adaptation of the book for a Broadway show in 1955 put Anne's story of her life in hiding on center stage. In 1957, the Anne Frank Foundation (*Anne Frank Stichting*) was set up under the leadership of Otto Frank to avoid the destruction of the 263 Prinsengracht house, which was opened on May 3, 1960, as a small museum. It would become from a best kept secret in town in the 1960s to one of Amsterdam's leading tourist attractions by the late 1990s when the "Maintenance and Future of the Anne Frank House" plan was implemented (Anne Frank House 1999; Westra 2000). The adjoining larger block area was restructured to meet the demand. The neighboring house Prinsengracht 265 became part of the museum tour, the corner building served for visitor services including an educational center to discuss and disseminate Anne Frank's ideals and humanistic values. Meanwhile the diary was translated into more than 60 languages and was read by tens of millions of mostly young people. By 2007, visitation to the Anne Frank House reached annually one million and the Anne Frank House launched a virtual reality version of the house (the *secret annex online*) to reduce capacity problems at the historic site. (Leopold & Amahorseija 2012; Hartmann 2013). In 2018, the Anne Frank House saw a re-routing of its visitation tour and the redesign of entrance situation to avoid the long lines of people waiting outside the House. The year 2019 saw a record visitation of 1,304,793. During the COVID pandemic the House was closed or when temporarily opened had the strict imposition of visitation guidance rules. It certainly helped that the Anne Frank House had an effective website, with visual tours available of the house with the secret annex. Website visits in 2021 were about 11 million, visits about 1.3 million to the Secret Annex. In the first post-pandemic year 2023, visitation reached the one million level again (1.208,112); website visits in this year were about 17.9 million, with 1.4 million visits to the Secret Annex (Anne Frank House Report 2024).

Anne Frank died with her sister, Margot, in the Bergen-Belsen concentration camp in March 1945. She is one of the most widely known victims of the Holocaust. Contemporary historians see her personal story as a "window to the Holocaust" (Young 1999). Anne Frank has worldwide a large following and an 'Anne Frank-Tourism' developed. Mostly young people follow in her footsteps, from her birthplace in Frankfurt and her years lived in Amsterdam (with an *Anne's Amsterdam* appt) to Bergen-Belsen (Hartmann 2017).

Figure 5.1 Anne Frank House from Prinsengracht 2018.

Source: courtesy Anne Frank Hause/Photographer: Cris Toala Olivares

There are two statues of Anne Frank in Amsterdam: one is located in the immediate proximity of the Anne Frank House on the Westermarkt Square (since 1977), and the other in front of the Merwedeplein apartment complex in South Amsterdam (since 2005), where the Frank Family lived during 1933/34–1942. The former Frank Family residence is not open to the public. The apartment now serves as the temporary home for persecuted writers who are given asylum in The Netherlands (Hartmann 2017).

5.2 The Verzets Dutch Resistance Museum (Plantage Kerklaan 61)

The Dutch Resistance Museum (Verzetsmuseum) was established in 1984/1985. The intention was to recreate the atmosphere during the war years. The museum exhibits displayed the major political, social and economic impacts the German occupation had in the Netherlands. In 1999, the museum was re-opened in enlarged form in the present site (Placius Building in Amsterdam's Plantation neighborhood) where it received originally about 50,000 visitors annually. Interest in the documentation of the resistance movement in the Netherlands, in particular in Amsterdam, has increased, and annual visitation reached over 100,000 in the late 2010s. Recently, the museum has been remodeled and its visitation has gone up again to the pre-pandemic level (2023: 113,730) according to Lisbeth V., CEO of the Verzetsmuseum. The majority of the visitors are Dutch, frequently children of school classes whose teachers have made the 1940–1945 period a 'must' topic in their curricula. The last remodel of the museum exhibits has contributed to an updated perspective of Amsterdam's Nazi German occupation, with the inclusion of more materials from the secretive operations of the resistance.

One new exhibit features resistance icon Hannie Schaft, a young student who became an active member of the Communist resistance in Haarlem, hiding and assisting Jews and carrying out several assassinations of infamous Nazi officials and well-known collaborators. The 'red-haired girl', as she was known to German forces, was caught on her bicycle carrying a gun and political pamphlets. On April 17, a few weeks before the end of the war, she was executed in the Dunes. After the liberation of the Netherlands in 1945, Schaft's body was dug up from a mass grave and later this year re-buried at the Honorary Cemetery in the seaside town of Bloemendaal, alongside hundreds of other resistance fighters. An updated documentary of her life, in the movie *To Die Beautiful*, this is what the New York Times article mentions (Moses 2023).

It is also worthwhile to mention a joint project of the Verzetsmsueum and the Anne Frank House: "Persecution and resistance in Amsterdam: Memories of World War Two – A walk from the Anne Frank House to the Dutch Resistance Museum" (Verzets Museum 2006). The walk features 33 stops at historic sites including the Homomonument on Keizersgracht honoring the hundreds of murdered (or persecuted) gay men in Amsterdam, the site of the illegal printing of the 'Het Parool' political press, the victims of the 'Groote Club' shooting on May 7, 1945, near the Dam Square, with 22 fatalities, the National Monument on Dam Square where annual ceremonies for the victims of the war are held on each 4th of May and the statue of the 'Dokworker' participating in the February 1941 strike.

The Homomonument, designed in the form of three pink triangles reaching into the Keizers Canal, commemorates all gay men and lesbians who have been persecuted because of their sexual orientation. Opened on September 5, 1987, it was the first monument in the world to commemorate gays and lesbians who were killed by the German Nazi regime. A similar unique political situation is celebrated with the 'De Dokers' statue on Jonas David Meierplein. It represents the local dock workers outraged by the harsh and fatal persecutions of Jews in their city. The February 1941 strike in Amsterdam was the only mass protest against the persecution of Jews in Europe.

5.3 The Jewish Cultural Quarter (Joods Cultureel Kwartier): the Jewish Museum (Joods Museum), the Portuguese Synagogue, the National Holocaust Museum and the Hollandsche Schouwburg

These four museums and/or historic sites are managed by the Jewish Cultural Quarter agency: Four centuries of Jewish culture spread over four venues, all within one square kilometer in the old Jewish Quarter of Amsterdam. The Jewish Museum – closed during the Nazi German occupation of the City – was re-opened in 1962. It displays the depth of Jewish culture in Amsterdam and showcases several synagogues from the 17th and 18th centuries. The Jewish Museum eventually comprised also exhibits on contemporary 20th-century history of Jewish culture in town including a section on the theme "Children and the Holocaust". The Holocaust for the whole state of the Netherlands will be explained and discussed in the new National Holocaust Museum opened in March 2024 (Lebovic 2024). The museum is located in a historic location, a former teacher training college whose kindergarten next door served as a deportation site for Jewish children. The main focus on the ground floor of the new Museum is on the local history and on the immense losses in the Amsterdam community. Though, they plan to cover also other sites of Dutch Jewish persecution like at concentration camps in Amersfoort and Kamp Vught, and the transit camp of Westerbork in Eastern Drenthe Province, from where 107,000 people were deported to the extermination camps in the East. The museum developed carefully designed educational programs, in particular for children in the upper years of junior school. The fourth site is the historic building of the Hollandsche Schouwburg, the municipal theater which functioned as the main deportation assembly point for Jews from Amsterdam in 1941/1942. The Schouwburg Memorial has been completely renovated and reopened in March 2024 together with the National Holocaust Museum. It will include an introductory film as well as portrait photos and audio recordings of victim's testimonies and memories of survivors.

The Schouwburg used to showcase a long list of the many deported individuals – now placed nearby at the National Holocaust Names Monument. The names of 102,000 Jewish victims from the Netherlands are shown in a monument

Figure 5.2a The National Holocaust Names Monument.

Figure 5.2b The name of Annelies Frank in the National Holocaust Names Monument.

1,550 square meters wide. Daniel Libeskind designed the memorial for the 102,000 people not given a proper burial. It was unveiled on September 19, 2021.

5.4 Archives and collections regarding the 1940–1945 war time period

Finally, it should be mentioned that a multitude of specific resources for the 1940–1945 time period are now available in the City of Amsterdam. Here, three major resources will be highlighted. The Anne Frank House has archives that comprise the press coverage of local and neighborhood events as well as some of the Frank Family collections from the Anne Frank Fonds in Basel, Switzerland. The Verzets Museum has many artifacts from the war years and bios of resistance fighters which are occasionally shown to the public in temporary exhibits.

The most intriguing archive is housed in the Netherlands State Institute for War Documentation. In a radio address on forbidden Radio Oranje (voice of the government in exile) Dutch citizens were asked to save documents about the war. Anne Frank was motivated by such a broadcast to get her diary ready for publication, and she started to rewrite some of her early entries. Thus, when Secretary Miep Gies saved Anne Frank's oeuvre after the arrest, she found not only the diaries' booklets but many loose pages. Otto Frank would eventually donate all the materials to the War Documentation institution. The documents were used eventually jointly with other pieces Anne wrote for the production of "The Critical Edition" published by the Netherlands State Institute for War Documentation in 1986/1989 (Frank, A. and Netherlands State Institute for War Documentation 1989), a "Definitive Edition" in 1995 and the revised critical edition of 851 pages in 2003. The latter was a precise comparison of the three versions that existed of Anne's diary: (a) her initial diary entries, (b) the revised sections of her diary and (c) Otto Frank's edited book version of 1947. As the fascination with Anne Frank's writings and with the time period 1940–1945 continues to persist, more surprise findings may well happen in the future.

Notes

1 Membership and support for the Dutch Nazi Party in Amsterdam were relatively small. Few people fell for the Nazi racial ideology that the Dutch as the Germans were Aryans. Though, acceptance of Nazi German doctrines and policies was substantially greater in the northeastern provinces, in particular in Friesland bordering Germany and sharing a joint dialect. The late professor Gregory Ashworth, who taught at the University of Groningen and studied local heritage culture and geography, once critically said that for each person in resistance another person was in collaboration. After liberation in 1945, 120,000 collaborators were put into prison in the Netherlands. Thirty-four members of the NSB, including fascist leader Anton Mussert, were executed.
2 Sources for the above information in the introductory section are the United States Holocaust Memorial Museum, the Anne Frank House, the Verzetsmuseum and The Holocaust Explained by the Wiener Holocaust Library.

References

Anne Frank House Annual Report. (1999). Amsterdam: Anne Frank House.
Anne Frank House Annual Report. (2024). Amsterdam: Anne Frank House.
Frank, A. (1947). *Het Achterhuis. Dagboekbrieven Van 14 Juni 1942–1 Augustus 1944 (Original Title: The Back Part of the House: Diary Entries June 14, 1942 – August 1, 1944*. Amsterdam: De Contactboekerij.
Frank, A. (1995) *The diary of a young girl: The definite edition*, Otto M. Frank and Mirjam Pressler (editors), Susan Massotty (translator). New York: Doubleday.
Frank, A. and Netherlands State Institute for War Documentation (1989). *The diary of Anne Frank: The critical edition*. New York: Doubleday.
Frank, A. and Netherlands State Institute for War Documentation (2003). *The diary of Anne Frank: The revised critical edition*. New York: Doubleday.

Hartmann, R. (2013). The Anne Frank House in Amsterdam: A museum and literary land-scape goes virtual reality. *Journalism and Mass Communication*, October, Vol. 3, No. 10, 625–644.

Hartmann, R. (2017). In the footsteps of Anne Frank, the Beatles and Vincent Van Gogh: A review and discussion of personal legacy trails. *Journal of Heritage Tourism*, published on line November 20, 2015, Vol. 12, No. 5, 463–473.

Lebovic, M. (2024). Netherlands' first Holocaust museum finally opens, with a side of anti-Israel protests. *The Times of Israel*, March 11, 2024.

Leopold, R. and Amahorseija, I. (2012). Interview with Ronald Leopold, Executive Director Anne Frank House, and Ita Amahorseija, Programme Manager for Digital Strategy at the Anne Frank House, August 28.

Moses, C. (2023). Overlooked no more: Hannie Schaft, resistance fighter during World War II. *New York Times*, July 7, 2023, Section B, p. 6.

Verzets Museum. (2006). *Persecution and resistance in Amsterdam – memories of World War Two: A walk from the Anne Frank House to the Dutch Resistance Museum*. Amsterdam: Verzetsmuseum (Dutch Resistance Museum), with collaboration of Hans Westra (Anne Frank Stichting).

Westra, H. (2000). *Anne Frank House: A museum with a story*. Amsterdam: Anne Frank Stichting.

Young, J. (1999) The Anne Frank House: An accessible window to the Holocaust. *Anne Frank Magazine*, 13.

Part II: Introduction

Remembering the Pacific War

Contrasting interpretations of the Pacific War events 1937–1945 and distinct forms of commemoration: the Japanese Greater East Asian War, Chinese resistance against the Japanese occupying forces and a Pacific wide engagement of the U.S. forces after the Pearl Harbor attack December 7, 1941

Rudi Hartmann

There are contrasting definitions and interpretations of the Pacific War events 1937–1945. Japanese, Chinese and American views have differed about the causes of the wars and the proceedings as well as about their outcomes. The Japanese Imperial powers, a colonial force in the Asia-Pacific region, saw the conflicts as an expression of a Greater East Asian war with the goal of assuring the leading role of the Japanese culture in this realm. The Chinese, both Communists and Nationalists,[1] focused on the resistance they practiced against the Japanese occupying forces; as such, they considered the events as an Anti-Japanese war that had started on July 7, 1937, with an incident at the Lu Gou Qiao Bridge south of Beijing. Millions of Chinese, both soldiers and civilians, were victims of this fight which ended with the defeat of the Japanese forces in 1945 and in an International Military Tribunal for the Far East (the 'Tokyo Trial') 1946–1948. After the Pearl Harbor attack on December 7, 1941, US forces were drawn into a Pacific-wide engagement which ended after two nuclear bombs on Hiroshima and Nagasaki with the 1945 surrender of the Japanese Imperial powers on the battleship MSS Missouri September 2, 1945. For Americans, the Pacific War was a theater or area of operations within the broader context of World War II.

How are the war events remembered by Japan, the People's Republic of China and the United States? What are leading or important memorial sites? Part II starts

DOI: 10.4324/9780367823795-9

with Chapter 6, with the focus on the July 7, 1937, incident on Lu Gou Qiao, also known as the Marco Polo Bridge. The confrontations between the Japanese and Chinese forces there are generally considered the beginning of the Second Sino-Japanese War. Japanese Imperial forces eventually occupied larger parts of China, but never the whole country. A Museum of the War of Chinese People's Resistance against Japanese Aggression was opened near the landmark Marco Polo Bridge in 1987, 50 years after the historic incident. The concluding Chapter 10 deals with Chinese (studying in the United States) visiting the memorial site at Hiroshima, where the Pacific War ended with the nuclear bombing in August, triggering the surrender of the Japanese forces on September 2, 1945. Here, the question is raised whether the Chinese and Japanese are willing to go past the horrible fatal war events between their nations, whether lasting peace can be restored in the Asia-Pacific realm.

The Hiroshima Peace Memorial Park is, besides the Yasukuni Jinja (Shinto shrine) with the Yushukan War Museum in Tokyo, the most recognized war memorial in the Japanese public. It honors the 400,000 plus direct and indirect (from long-term effects of exposure to radiation) victims after the dropping of the nuclear bomb on the city center. More than 20 individual memorials and monuments are part of the memorial park. The museum has a main exhibit that shows the horrors of nuclear war. The museum exhibit starts with the day of the bombing (August 6, 1945). Smaller additional exhibits that tell the reasons for the bombing and the history of the war in general were installed outside the main exhibit later in the re-organization of the museum. While 'Hiroshima' shows the Japanese at last as victims of the war, the Yasukuni shrine and the Yushukan War Museum tell a different story, that of pride in the leadership and in the soldiers who carried the war through a long period of expansion of the Japanese controlled territory in Asia-Pacific and again through the years of retreat after 1942. About 2.5 million Japanese war dead, including 13 A-Class war criminals, are commemorated at the Yasukuni shrine. Actually, they were apotheosized as they were elevated to god-like personae after their faithful service to the Japanese emperor. The war museum on the site gives solemn tribute to them. The war interpretations at Yasukuni and Yukushan have led to political conflicts within Japan and with Asian neighbors, like China, who were tragically affected by the actions (and atrocities) of the Japanese Imperial powers. The Yukushan War Museum and the Chiran Peace Museum honor specifically the Kamikaze pilots, the 3,000 airmen who died in suicide flights between October 1944 and September 1945, as discussed in Chapter 9. Here, the question is raised whether visitation of the sites that discuss these specific war issues in a broader contemporary sense can contribute to a better mutual understanding and to peace.

Similar issues are discussed in Chapter 8 on the Yamato Museum in Kure, a naval port town next to Hiroshima. While Hiroshima became a peace memorial city (see, for instance, Katayanagi and Kawano 2023), Kure retained largely the legacies of its war industry. It was in Kure where a military shipyard evolved and flourished, where the large Yamato battleship was built. The museum opened in 2005, with a 1/10 model of the battleship which sank in 1944. Interestingly enough, contemporary Japanese culture has taken up the fate of the ship and of its men in novels, films and comic strips. Both the continuation and the disconnect with its war heritage are present here. A multitude of new war related museums in Japan have gone through these changes from a peace-motivated museum to more war-focused exhibits and back to peace.

Two events and their anniversaries are routinely remembered by the American public: 'Pearl Harbor Day', December 7 (1941), and the end of World War II, in Europe on May 8, 1945, and in the Pacific on September 2, 1945. The memorial at Pearl Harbor is one of the most visited heritage sites in the United States and is managed by the National Park Service. More than one million people annually visit the memorial site in Honolulu, Hawaii. It was built distinctly into the harbor halfway underwater to honor the 2,000 plus victims of the Japanese surprise attack. The battle over the Pacific realm in 1942–1945 saw some of the bloodiest campaigns: at Guadalcanal, at Midway, at Iwo Jima and at Okinawa. In Iwo Jima, for instance, more than 7,000 Japanese soldiers and more than 2,000 marines died. After more than 80 days of combat, Okinawa was secured by the American troops, with even higher numbers of fatalities on both sides (Dunnigan and Nofi 1995; Gailey 1996). The raising of the American flag at the remote island of Iwo Jima (700 miles East of Tokyo) has become a memorial theme in itself, with a memorial site outside Arlington National Cemetery and several other places in the United States.

No discussion of the memorial sites of the Pacific War would be complete without a reflection on the sites where the U.S. government incarcerated Japanese American individuals and families during World War II. In popular media, this history often begins with the U.S. reaction to the bombing of Pearl Harbor by the Japanese Army, but the incarceration is best contextualized by deep-seated racism and exclusionary U.S. policy directed at people of Asian descent in the U.S. (including the) Chinese Exclusion Act of 1882.

(Peterson and Clark 2024)

Chapter 7 reviews the history of two of the ten detention camps, Manzanar and Amache, of the War Relocation Authority (WRA), where over 120,000 Japanese Americans were removed from their home areas in the American West and incarcerated till the Pacific War ended in September 1945. It was not until the 1960s/1970s that a formal apology and some modest remuneration were given to the former internees for their unjust treatment. By the 1990s/2000s all the camps became historic landmarks and destinations for personal, family and community pilgrimages. The war-dead of the Japanese-Americans are remembered in the cemetery of the camps; in Amache, Colorado, the names of 31 young men are listed. Each May, during the pilgrimage, the lives of the young men who volunteered to fight in Europe for their nation and home country are commemorated.

Over 20 million Chinese, soldiers and civilians, died during the eight years of Japanese occupation, 1937–1945 (Hsiung and Levine 1992; Mitter 2013). Several memorial sites recognize the victims of the war. The memorial sites, including the one at the prominent Anti-Japanese War museum at Lu Gou Qiao, not only precisely reconstruct the war events but also display the war brutalities of the Japanese Imperial army, including the 'rape of Nanjing' in December 1937/ January 1938 (Chang 1997; Brook 2000). Visits to Lu Gou Qiao have seen the emergence of 'dark tourism', a personal curiosity in the war atrocities, and of 'red tourism' in the 2000s/2010s, as a form of national education regarding the historic achievements of the Red Army, as discussed in Chapter 6.

It should be mentioned here that there were other nations and peoples in the Asia-Pacific realm which were affected by the war events like the Philippines, Indonesia and New Guinea. Probably, the most severely punished nation was Korea, which had to endure decades of Japanese occupation and colonization. Hiroshima Peace Memorial Park has a Korean monument to honor about 45,000 Korean victims who lived and worked in the city during the dropping of the nuclear bomb. Another large population segment of victimization were Korean and Chinese females who were forced by the Japanese Imperial army to serve as "comfort women" (Qiu, Su and Chen 2014; Ramaj 2022; Soh 2009). As Japan continued military expansion before and during World War II the military turned to local populations in the occupied territories. It is estimated that 50,000 to 200,000 women and girls were abducted and coerced to service as sex slaves in the established 'comfort stations'. In the 2000s/2010s, memorial sites and memorial statues were established in Seoul, South Korea, at Nanjing, China and at several cities including San Francisco in the United States.

A multitude of atrocities committed by the Japanese Imperial forces in the 1930s and 1940s continue to put a strain on current relations between Japan and the nations in East and Southeast Asia. Rarely, Japanese officials have formally expressed and given concrete apologies for the horrific actions of yesteryear, as, for instance, done by Japanese Prime Minister Tomichi Murayama, who offered his apology to the comfort women from Korea and through his visit and words at the Lu Gou Qiao Memorial Museum in 1995. The inherent misgivings over the sins of the past have persisted into today's world, and the strategies and practices of how to overcome them are discussed in Chapter 10.

Note

1 In the 1930s/1940s, the CCP (China's Communist Party), under the leadership of Mao Zedong, and the Nationalists (Kuomintang), under the leadership of Chiang Kai-shek, were involved in a bitter, long-lasting Civil War; at times, though, Communists and Nationalists collaborated on their fight against the Japanese which occupied parts of China. In 1948, Mao pronounced the People's Republic of China, whereas Chiang established an autocratic regime in Taiwan. Chiang died in 1975, Mao in 1976. In the 1980s/1990s/2000s reform period, the Nationalists' contribution in the Anti-Japanese War was more and more appraised by Communist China (Mitter 2013). ii See Hartmann 2013, 2015.

References

Brook, T. (2000) *Documents on the Rape of Nanking*. Ann Arbor: University of Michigan Press.

Chang, I. (1997). *The Rape of Nanking: The Forgotten Holocaust of World War II*. New York: Basic Books.

Dunnigan, J. & Nofi, A. (1995). *Victory at Sea: World War II in the Pacific*. New York: William Morrow and Company.

Gailey, H. (1996). *War in the Pacific: From Pearl Harbor to Tokyo Bay*. San Francisco: Presidio Press, 1996.

Hartmann, R. (2013). *Tourism to Heritage Sites with a Controversial History: A Case Study of Eight Historic Sites Associated with the Pacific War 1937–1945*. Abstract of paper presented at the International Conference on Tourism Landscape and Tourism in Marginal Areas, Kanas, Xinjiang, China, September 23.

Hartmann, R. (2015). *Tourism to Heritage Sites with a Controversial History: The Commemoration of War and Peace at Sites of the Pacific War*. Invited Taoyaka Lecture, Hiroshima University, Higashihiroshima, Japan, May 22.

Hsiung, J. & Levine, S. (1992). *China's Bitter Victory: The War with Japan, 1937–1947*. New York: M.E. Sharpe. Reprinted by Routledge 2015.

Katayanagi, M. & N. Kawano (2023). Reconstructing Hiroshima as A Peace Memorial City: Local Agency and Identity-Making in Peacebuilding. *War & Society*, 1–19, School of Humanities and Social Sciences, UNSW Canberra.

Mitter, R. (2013). *Forgotten Ally – China's World War II, 1937–1945*. Boston: Houghton Mifflin Harcourt.

Peterson, W. & Clark, B. (2024). Chapter 7.

Qiu, P., Su, Z. & Chen, L. (2014). *Chinese Comfort Women: Testimonies from Imperial Japan's Sex Slaves*. Oxford: Oxford University Press.

Ramaj, K. (2022). The 2015 South Korean-Japanese Agreement on 'Comfort Women': A Critical Analysis. *International Criminal Law Review*, 22 (3), 475–509.

Soh, S. (2009). *The Comfort Women: Sexual Violence and Postcolonial Memory in Korea and Japan*. Chicago: University of Chicago Press.

ISSN: (Print) (Online) Journal homepage: https://www.tandfonline.com/loi/rjht20

6 Tourism to Lu Gou Qiao

Enduring scenic qualities of a landmark bridge and a difficult legacy of a conflict site (*Journal of Heritage Tourism*, Vol. 16, 2021, Issue 6, 705–715)

Rudi Hartmann and Ming Ming Su

Introduction

Lu Gou Qiao, a historic twelfth century bridge in Southwest Beijing also known as Marco Polo Bridge, was the site of an incident on 7 July 1937 which marked the beginnings of the Second Sino-Japanese War with twenty million or more fatalities (mostly among Chinese civilians) during 1937 and 1945 (Hsiung & Levine, 1992; Mitter, 2013). In 1987, fifty years after the incident that led to the war, a Museum of the War of Chinese People's Resistance against Japanese Aggression was established in the neighboring Town of Wanping. Since its inception in 1987, the museum has drawn more than 33 million visitors till the end of 2018 (1987–2018) (http://www.1937china.com/kzjng/ views/bgjs/bgjs.html, retrieved on 10 April 2019). The present study examines visitation to the bridge site and to the museum.

A main research question is what motivates tourists to visit the museum and what are the primary reasons for seeing the landmark bridge, an attraction in itself with about 500 stone lion sculptures along an old cobblestone pathway. Further, the assumption is tested whether visits to the two sites are combined.

It is argued here that the two adjoining sites – about 1 km physically apart – are not only thema- tically connected but are also complementary in nature. Our studies and empirical research at the museum and the bridge show that visits to both sites are most often organized and experienced together. For some of the visitors, the two sites have attributes of a difficult and dark heritage. Thus, the notion of 'dark tourism' to Lu Gou Qiao and the Museum of the War of Chinese People's Resistance against Japanese Aggression will be explored. Further, 'red tourism,' heritage tourism to prominent sites of the Communist movement, is examined. Patriotic education according to the agenda and guidelines of the Communist Party of China (CPC) plays a major role at the Museum.

The Role of dark tourism and red tourism in the visit of heritage sites

Both 'dark tourism' and 'red tourism' can be considered distinct forms and expressions of heritage tourism, a term that has largely replaced the concept of 'cultural

DOI: 10.4324/9780367823795-10

tourism.' Tourism researchers and geographers found the traditional cultural tourism term too narrow and elitist. By contrast, heritage can be widely understood as all the tangible and intangible features of the past we wish to preserve in present day society. Thus, heritage is selective and its formation is part of a dynamic process (Hartmann, 2014, pp. 168–170; Timothy & Boyd, 2003; Williams, 2009, pp. 236–257).

Lu Gou Qiao is among the many prominent heritage sites in China and East Asia discussed in the literature (see, for instance, Prideaux et al., 2008). In particular, Lu Gou Bridge marks a well recognized cultural heritage site. Outstanding examples of architecture like the Great Wall of China (Su & Wall, 2012) have attracted millions of visitors. What makes the Lu Gou Qiao Bridge and the Museum heritage sites special is the unique appeal of the destination for 'dark tourism' and 'red tourism.'

Both concepts are fairly new to the scholarly debate and in the media. By the mid and late 2000s, the terms dark tourism and red tourism were firmly established. Both terms have a regional origin. Whereas dark tourism was coined in and disseminated from Southern Scotland and Northern England, with first articles and books by UK authors (Foley & Lennon, 1996; Hartmann et al., 2018; Lennon & Foley, 2000; Stone, 2006; Stone & Sharpley, 2008), the notion of red tourism came out of China. This politically defined term was first proposed by Chinese Communist authorities in 2005 and 2006, with nationwide implementation in the following years (Hung, 2018; Li et al., 2010; Wall & Zhao, 2009).

What is dark tourism? Philip Stone, founder of the Dark Tourism Forum and later of the Institute of Dark Tourism Studies, defined it with the following words: 'the act of travel and visitation to sites, attractions and exhibitions which have real or recreated death, suffering or the seemingly macabre as a main theme' (2005). A multitude of studies which discussed the theory and practice of dark tourism (Sharpley & Stone, 2009; Stone, 2006; Stone & Sharpley, 2008) followed; they ranged from visits to WW I battlefield sites and memorial sites of the Holocaust to Dracula Castle and favela tours. Efforts have been made to develop the dark tourism (and its parallel thanatourism) approach to a globally appealing research perspective characterizing a discrete form of traveling beyond mere special interest tourism within the wider heritage tourism field (Light, 2017; Seaton, 2009; Stone, 2013; Stone et al., 2018). It is noteworthy that also several Chinese tourism researchers examined dark tourism sites in the People's Republic of China, specifically for the visitation of earthquake memorial sites, such as the 1976 Tangshan and 2008 Wenchuan earth quake sites, and of the Nanjing Massacre site (Chen & Xu, 2017, 2018; Tang, 2014a, 2014b, 2018a, 2018b; Wang & Luo, 2018; Zheng et al., 2018). Lu Gou Qiao represents a site with a 'dark' connotation of the place. Visitors to the Bridge and the Museum are confronted with narratives of death and tragedy the Chinese people endured in their fight against Japanese aggression.

Red tourism is the popularization of a term that describes state sponsored and organized tourism to sites that are closely tied to the Communist movement. Leading members of the Communist Party of China (CPC) felt in the early and mid-2000s that there was a need to strengthen the Communist heritage in China, in particular among the young generation. The promoted Communist heritage sites range from the meeting place where the Communist Party of China was founded in Shanghai during its first National Congress in 1921 and the sites where the legendary Long March played out in the 1930s to the birthplace of Mao Tse Dong in Shaoshan, Hunan Province, and his mausoleum on Tiananmen Square in Beijing. Five locations received the designation of a Revolutionary Sacred Site (Hung, 2018). In general, there has been a tendency to place and develop red tourism sites in rural parts of China, as an economic development strategy. Since the launching and implementation of a National Red Tourism Plan in 2008/09 twelve significant red tourism regions have been identified, with 30 red tourism itineraries and about a hundred classic red tourism sites in virtually all provinces and autonomous regions of China. The selected places have the recognized heritage status of Communist historic sites. The main purpose and agenda of red tourism sites is to conduct education in patriotism and the revolutionary tradition (Hung, 2018; Li et al., 2010; Wall & Zhao, 2009). The Beijing Municipality has seven sites including Lu Gou Qiao. The Museum is an officially recognized red tourism site. The Bridge, as a landmark cultural heritage site, has also seen official activities of the Party, like the swearing in of new members of the CPC.

Research setting: Lu Gou Xiao bridge and the museum of the War of Chinese People's Resistance against Japanese Aggression in the Town of Wanping

The research was prepared with interviews of managers at the Museum (Assistant Director of Exhi- bition Department, Director of the Department of Historical Studies) and with the Director of the Lu Gou Bridge Cultural Tourism District Office (which is part of the Lu Gou Bridge Department of Cultural Relics Protection and the Beijing Municipal Bureau of Cultural Relics). The Museum is a vice- ministerial level institution under jurisdiction of the Beijing municipal Party Committee Propaganda Department, with responsibility for the organization of red tourism, and has 137 full-time employees. Until 2008, an entrance fee of 15 Yuan was charged. Since then, visits to the Museum are free of charge. The Lu Gou Bridge Cultural Tourism Zone employs 120–130 persons. The Cultural Tourism Zone contains Lu Gou Bridge, the walled Wanping City and outside the walls a Sculpture Garden of the Anti-Japanese War (see Figure 6.1). The Cultural Tourism Zone is fully funded by the local government (Beijing Fengtai District). Lu Gou Bridge charges a 20 Yuan entrance fee. According to Liu Hui, the Director of Lu Gou Bridge Cultural Tourism Zone, the Bridge is not a red tourism site, but a historical and cultural tourism destination. Though, the Bridge has seen activities organized by Party committees as a form of patriotism education.

A careful examination of the websites of the Museum and of the Bridge showed that both sites receive significant visitation. They give basic quantitative data regarding the number of visitors to the two sites as well as about information about the major policies and practices of the management of the Museum and the Bridge. Since 2011 the museum has received on an average 600 thousand Visitors annually with one year (2014) surpassing one million. Visitation at the Bridge was lower, with about 250 thousand visitors annually in the years 2011–2013. In 2014 a peak

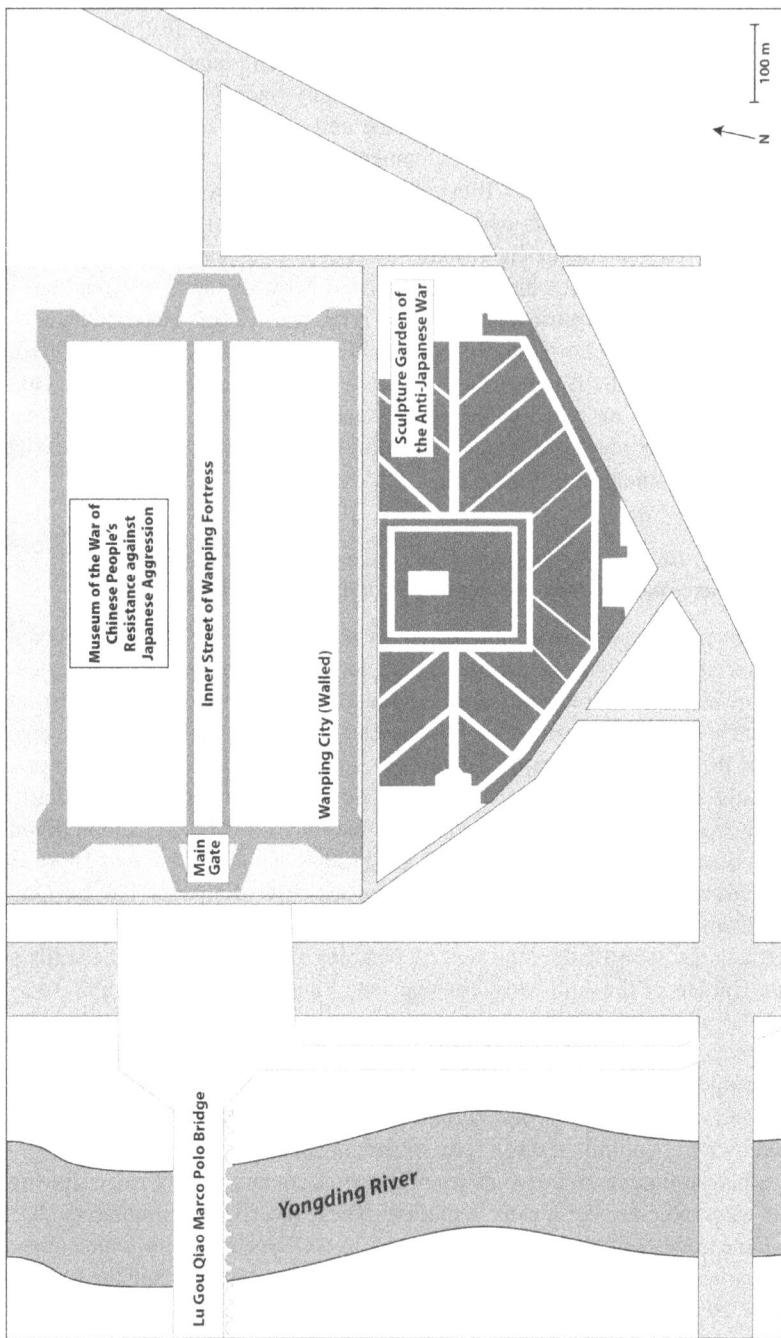

Figure 6.1 The Relative Location of Lu Gou Bridge and the Museum (redrawn by Peter Anthamatten from Lu Gou Bridge Cultural Zone 2020 Floor Plan).

number of 622 thousand visitors was recorded. Lu Gou Bridge is a semi-enclosed tourism area as it has to provide access to 400 households who live across the Bridge on Xiao Yue Island. Thus, the Bridge is kept open year-round, with guardians in charge of the access to the bridge site.

The Lu Gou Qiao Bridge as a historic tourist attraction

Original construction of Lu Gou Qiao occurred between 1189 and 1192, during the Jin Dynasty. It was built over the Yongding River – formerly known as Lu Gou ditch – 15 km southwest of the Beij- ing city center. The bridge was the main access to the capital area for travelers coming from the North China Plain. The bridge has a complex and elaborate history; it has attained landmark signifi- cance since the thirteenth century. It is the oldest existing multi-arched stone bridge in the Beijing area. Venetian merchant and traveler Marco Polo highlights Lu Gou Qiao in his diary with the fol- lowing words (in two different translations): 'This river is crossed by a magnificent stone bridge. There is not a bridge in the world to compare with it . . .' (Latham, 1958, p. 163)

> Over the river there is a most beautiful stone bridge, truly the finest in the world, and without equal. It is no less than 300 paces long and eight paces broad, so that as many as ten horsemen can cross riding abreast. It has twenty-four arches, and twenty-three pillars on which the arches stand. It is all of grey marble, most excellently worked and put together. Along each side there is a parapet of marble slabs and columns. The way that leads up to the bridge is rather broader at the bottom than at the top, but once you reach the top, the bridge is of one width all the way, as if it had been drawn with a line. At the beginning of the bridge, there stands a very large and tall marble column, resting upon a marble tortoise, with a big marble lion at the foot; another very beautiful lion, big and well-made, lies on the top of the column. So that it is a truly splendid sight.
>
> (Kin, 1981, p. 62)

Subsequently, Lu Gou Qiao has been known by Westerners as 'Marco Polo Bridge.' In a famous painting of the Yuan Dynasty ('Transporting logs across Lukouch- iao'), the bridge is depicted in a way Marco Polo may have seen it (Kin, 1981, pp. 56–57). The painting renders a view of the commerce and trade that took place over the bridge and at the banks of the river. In his travel diaries Marco Polo fre- quently included observations regarding the economic geography prevailing in the Yuan Dynasty and in particular during then governing Great Khan. During the reign of Emperor Kangxi, from the Qing Dynasty, the bridge was rebuilt. Lu Gou Qiao is 266.5 meters in length and 9.3 meters in width supported on 11 piers over 10 arches. At the entrance side of the bridge stand two steles, one denoting the year it was reconstructed (during the 37th year of the reign of Emperor Kangxi, in 1698 AD), and the other one in Chinese calligraphy reciting the poem and famous description of the bridge by Emperor Qianlong (1711–1799): 'the morning moon of Lugou.' Ever since, it has been known as a scenic spot in Beijing (Hou, 2015).

What makes Lu Gou Qiao so remarkable are the large number of unique stone lion statues on both sides of the bridge. Most of the lion statues have additional smaller lions included in the composition such as a second lion hiding on the head, on the back or under the belly or on paws of each of the big lions. Close to 500 stone lions are now posted on the white marble pillars. It is said that there were orig- inally 627 lions. Most lion statues date from the Ming (1368–1644) and Qing (1644–1911) dynasties. Some are from the earlier Yuan Dynasty (1271–1368, when Marco Polo visited) while the few lions dat- ing from the Jin Dynasty (1115–1234) are now quite rare (Jiunn, 2011; Kelly, 2017; People's Daily, 2017). The water level of the Yongding River varies as its waters are diverted to different areas of Beijing during the year. At times, there is no water running under the bridge; at other times, it has a sub-stantial flow of waters which allows boating and recreation on the Yongding.

The Museum as a tourist attraction in neighboring Wanping

The fiftieth anniversary of the Lu Gou Qiao Incident in 1987 was the year chosen for the official commemoration of the conflict and war events in form of a new museum and hall of fame in Wanping.

There are several names denoting the incident and battle between the Chinese and Japanese troops: Marco Polo Bridge Battle or Incident in the international literature, the Lu Gou Bridge or Lu Gou Qiao Incident with the given local name of the bridge, and the July 7th Incident (or 7 7 Incident) in the Chinese and Korean literature. It is commonly agreed that the incident marked the beginnings of the Second Sino-Japanese War 1937–1945.

The *Museum of the War of Chinese People's Resistance against Japanese Aggression* was formally opened on 7 July 1987. The main functions of the museum are to provide an educational basis to carry out patriotic education, to encourage the study of the history of the Anti-Japanese War and to tell a story of the war of resistance in narratives and songs in a national spirit auditorium. The museum underwent several phases of reconstruction and modernization, after 1987 the Phase Two renovation in 2000, the Phase Three renovation in 2005 and the Phase Four renovation in 2015. New sections were added on including the con-struction of the Taiwan Anti-Japanese War exhibit. The museum now has a display area of more than 6,700 square meters with rich collections of exhibits to show the history and huge sacrifice of the Chinese people's resistance against Japanese aggression. It also discloses the monstrous crimes the Japanese invaders committed during their aggression against China from 1937 to 1945.

The management of the museum plans to develop more interactive areas includ-ing exhibits which provide more activities for the younger visitors to the museum. Each year, there are four to five special exhibits which are covered in the local and national media. Finally, the museum gives access to virtual exhibits on its web site.

The peak tourist season for visits to the museum ranges from the Ching Ming (April) Festival to National Day in October. Important days of visitation are 7/1 (The Party's Day), 7/7 (The 7 7 Incident Anniversary Day), 8/15 (The Japanese Surrender Day), 9/3 (The Chinese Anti-Japanese War Victory Day), 10/1 (The National Day), and the 12/13 (The Nanjing Massacre Victims National Memorial Day). The museum is frequently visited by VIPs on one of the above days, most prominently on the 7/7 Anniversary Day, when Chairman Xi Jinping visited in 2014 and 2015.

The museum is visited by many school, party and local groups. They are assisted by more than 20 interpreters – as tour guides – during the peak season. The interpreters are volunteers and usually retirees. They go through rigorous training. The museum is also equipped with more than 100 voice navigation devices which can be rented by the visitors. Devices with the following four languages are available: Chinese, English, Japanese and Russian.

Research design

The field research at Lu Gou Bridge and the Museum was conducted from mid-October to the end of November 2017. A total of 267 questionnaires were collected at the Bridge and 265 at the Museum. Questions were asked about the respective duration of the visits at the sites and whether the sites were visited for the first time or were visited before. The respondents also gave answers as to which site (the Bridge and the lion sculptures versus the Museum) they found most impressive. Likert scale questions gave details about three types of information: the respondents' motivations to come to the sites, their perceptions at the sites and their evaluations of the site visits. Finally, the empirical research provided information about the demographics of the visitors (Figures 6.2 and 6.3).

Findings

As expected, a major result of the survey was that the Museum and the Bridge site were visited jointly by a large majority (72%). Which of the sites was most impressive? After the visits to the Museum and to the Bridge respondents stated that the museum visit was slightly more impressive (45% museum, 60% jointly museum and bridge/lions versus 37% bridge/lions, 47% jointly bridge/lions and museum). Duration of the visit was longer at the Bridge. Respondents stated that they spent two to four hours there (45%), less than two hours (45%), with another 8% spending 6–8 h at the bridge site, whereas a majority (56%) spent two hours or less in the museum; only 36% and 5% spent 2–4 h and 6–8 h respectively there. One of the explanations for the shorter visits at the museum might be the many more guided tours at the Museum (50% of the total visitors). Satisfaction with interpretation at the museum including tour guided interpretation was rated very high on the Likert scale (4.17). The bridge site was better known from previous visits. 65% of the respondents visited the Bridge for the first time while the other third stated that they had been there before (16% a second time, 8% a third time and 20% four to ten

Figure 6.2 The Lu Gou Xiao Cultural Landmark Bridge (photograph by Ming Ming Su).

Figure 6.3 Group Visitors at the Bridge (photograph by Rudi Hartmann).

times). In the case of the museum, first-time visits made up 73%; second and third time visits accounted jointly for 19%.

The highest motivations to visit the two sites were first the desire to review the history of anti- Japanese war and secondly patriotism education (with about 4.4% each on the Likert scale). The interest in war related information was also rated very high (4.3). Three motivations for the visit to the Bridge, the high reputation of the bridge, an appreciation of Chinese traditional architecture and a desire to better understand Chinese traditional culture and art, were rated in the range of 4.1 on the Likert scale

The highest ranked perceptions of the sites were associated with patriotic educa- tion and feeling more patriotic after visiting Lu Gou Bridge, the Museum (in the range of 4.4 on the Likert scale). High ratings were given for visiting the sites (both the Bridge and the Museum) and contributed to a pride in being Chinese. Visitors felt the hardship of the events of the times. Slightly lower ranked were the percep- tions of visiting Lu Gou Bridge as helping the visitors to better understand tradi- tional Chinese culture as well as Chinese architecture (4.2). Lu Gou Bridge was considered both a red tourism site (4.4) and a cultural tourism site (4.15). Slightly lower ratings were given to the statements that the Museum is a 'must see' site in Beijing (4.17) and that Lu Gou Bridge is a 'must see' site in Beijing (4.15). The rated perceptions of 'I think of the stone lions when Lu Gou Bridge is mentioned' (4.2) and of 'I think of the famous 'moon on the Lu Gou Bridge' as one of the eight famous views of Old Beijing when Lu Gou Bridge is mentioned' (4.0) were still above 4 on the Likert scale.

High levels of satisfaction were associated with the architecture, landscape and protection of cultural heritage for the Bridge (around 4.1 and 4.0) whereas the Museum's overall satisfaction and with its facilities and exhibitions ranked even higher, in the 4.3 and 4.2 range. The ticket price at the Bridge was considered less satisfactory (3.1), with the level of crowding there getting the lowest rank- ings (2.58). Crowding was also experienced at the Museum (3.16).

Thematic connectedness and complementarity of the two sites

We argue that the Bridge and Museum are thematically connected and that they are complementary in nature. Both sites provide a focus on a major chapter or date in the history of China and foremost the events that occurred on or near the bridge the 7th and 8th of July 1937. The Lu Gou Qiao incident and battle between the Chinese and Japanese troops are shown in great detail in one of the display areas of the museum. The museum exhibit reconstructs the situation at the outbreak of the war as well as the moving battle lines. The Japanese Imperial Army claimed a missing soldier as reason for occupying parts of Wanping and the bridge area. Eventually, the conflict gave way to the full-fledged invasion of China by Japan in the follow- ing months and years. The historic events can be reviewed and personally revisited both on the bridge and in the museum.

The two sites also complement each other. While the focus in the museum exhibit is on the accurate historical reconstruction of the incident that led to the Second

Sino-Japanese war, the bridge gives the topographical background to the conflict. Moreover, the bridge represents an outstanding example of art and architecture of the Chinese civilization. It is an illustration of why Chinese people could be or should be proud of their cultural achievements and contributions to world civilization. The museum, on the other hand, is able to provide a perspective on Japanese aggression, first with the occupation of the Northeast provinces ('Manchuria') of China 1931 and in the later war activities 1937–1945. The Wanping Museum of the War of Chinese People's Resistance against Japanese Aggression is the leading site in China to give a full account of the unfolding of the Second Sino-Japanese War. The museum exhibits include horrific events such as the bombing of Shanghai and of the Nanjing Massacre as well as give information about the end and outcome of the war, with China's victory and the return to peace in 1945.

Lu Gou Qiao as a site with a dark and difficult heritage

In the mid and late 1990s, several new concepts entered the debate in contemporary tourism studies: dark tourism, thanatourism and dissonance in heritage management (Foley & Lennon, 1996; Hartmann, 2014; Lennon & Foley, 2000; Tunbridge & Ashworth, 1996). These new concepts and approaches have contributed to the formation of new research traditions now widely used in scholarly tourism studies and in the media. As discussed earlier, a wealth of literature, in particular in the 'dark tourism' studies field, is available (Light, 2017; Stone, 2013; Stone et al., 2018). There have also been dark tourism studies in China. Most recently, the Nanjing Massacre site, with a large and heavily fre- quented memorial site in Nanjing, was examined from a dark tourism perspective: 'The inner struggle of visiting "dark tourism" sites' (Zheng et al., 2018). The question regarding tourism to Lu Gou Qiao bridge and the Wanping museum is whether some of the visitors are attracted to the sites because of the many and grim fatalities that occurred 1937–1945. They are referred to and/or shown at both Lu Gou Qiao memorial sites. It can be argued that touring the two sites provides a close-up of death and dis- aster in the Chinese historical context. Visitors to the museum site, in particular, are confronted with atrocities and the fragility of life during the war years. With a growing number of individual visitors in recent years, it is most likely that some of them recognize the darker side of humanity and consume aspects of a 'dark tourism' experience during their Lu Gou Qiao visits. Survey results such as the high interest in war-related information (4.3 on the Likert scale) including massacres and horrific events confirm our dark tourism assumptions.

It should be mentioned that the number of Japanese visitors has gradually and consistently diminished over the past years. While two Japanese Prime Ministers, Tomichi Murayama in 1995 and Junichiro Koizumi in 2001, visited the museum and subsequently many of the Japanese visitors laid flowers and/or expressed thoughts of regrets and remorse, this trend has faded. The Sino-Japanese relations having recently become more contentious and divisive. International visitors from the U.S. and Western Europe represent only a very small portion (since 2009 below 3%) of the visiting public to Lu Gou Qiao. It remains to be seen whether the relatively low interest in the Lu Gou Qiao landmark bridge as well as the incident and the following 1937–45 war events mark a permanent situation or is a passing trend.

Lu Gou Qiao as a red tourism site

Research results indicate that visitors to the Museum are highly motivated by the patriotism education opportunity (Mean = 4.384) provided by the Museum to revisit the history of the Anti-Japanese War of China (Mean = 4.416). Most visitors acknowledge that their visiting experiences enhanced their understanding of the history of the Chinese Anti-Japanese War and that they have become more patriotic after their visits (Mean = 4.360).

In addition, the majority of visitors acknowledge that they acquired a deeper understanding of the cruelty of the war (Mean = 4.358), thus treasure more of the peace of today (Mean = 4.358). Some- how, visiting the Museum also give visitors a heavy heart (Mean = 4.165) after witnessing the cruelty of the war and hardship of the Chinese people at that time.

Therefore, research results revealed that the visiting experiences to the Museum reinforce the collective memory of the Chinese people, strengthen their national identity, enlighten their patriotic emotion, and encourage them to treasure the peace today and contribute to the future of China.

Conclusions

Our studies show that the Lu Gou Qiao 'Marco Polo' Bridge and the Museum of the War of Chinese People's Resistance against Japanese Aggression in nearby Wanping are most often visited jointly. The combined experience of the two places is characterized by its thematic connectedness, as the bridge site formed the back-drop to the incident where the Second Sino-Japanese War 1937 started and as the museum reconstructs the beginnings, the course and end of the war in great detail. While the bridge visitor is able to learn about the topography of the incident setting, the museum focuses on the accuracy in the presentation of the historic events that followed. Visiting both sites is also marked by complementarity. While the landmark bridge represents a heritage site which displays Chinese achievements in art and architecture during past centuries and marks an outstanding contribution to world civilization, the museum provides a narrative that tells the story of the very beginnings of modern-day China. The guided, informative museum experience serves for the patriotic education needs at a red tourism site complemented by a high motivation of the visitors to learn about China's past. The stated interest in grim and horrific war related information is an indication of an evolving 'dark tourism' to both sites, in particular among the individually visiting people. Thus, important forms of motivation for visiting Lu Gou Qiao is participation in 'red tourism' and an increasing openness to 'dark tourism.'

Lu Gou Qiao may still be an underrated tourist attraction hardly visited by foreigners including Japanese. An analysis of the visitation trends has clearly seen a decrease in Japanese visitors which can be attributed to the rising tensions (and expressions of nationalism) between Japan and China in recent years. Expressions of feeling good about being Chinese after the visit makes Lu Gou Qiao a heritage site which has remained largely a domestic tourism destination. The perceived difficult legacy of the July 7th, 1937 incident and conflict of how China resisted and successfully overcame Japanese aggression contributed to the establishment of a distinctly dark and red tourism site in the Greater Beijing Metro Area.

Disclosure statement

No potential conflict of interest was reported by the author(s).

Notes on contributors

Dr. Rudi Hartmann is a Professor C/T at the Department of Geography and Environmental Sciences, University of Colorado Denver where he has taught geography and tourism planning since 1992. A long- time research interest of his is the study of heritage tourism, and of memorials sites of the Holocaust and of the Pacific War. He is co-editor of the Palgrave Handbook of Dark Tourism Studies.

Dr. Ming Ming Su is an Associate Professor at the School of Environment and Natural Resources, Renmin University of China, Beijing. She holds degrees from the University of Waterloo in Canada and Tsinghua University of China. Her research focuses on heritage management, tourism impacts, tourism and community relations, tourism at protected areas, and tourism issues in China.

References

Chen, S., & Xu, H. (2017, April 7). *The globalization of commemoration? Changing natural disaster memorial landscapes in China.* Paper presented at the Annual Meeting of the Association of American Geographer, Las Vegas.

Chen, S., & Xu, H. (2018). From fighting against death to commemorating the dead at Tangshan earthquake heritage sites. *Journal of Tourism and Cultural Change, 16*(5), 552–573. https://doi.org/10.1080/14766825.2017.1359281

Foley, M., & Lennon, J. (1996). JFK and dark tourism: A fascination with assassination. *International Journal of Heritage Studies, 2*(4), 198–211. https://doi.org/10.1080/13527259608722175

Hartmann, R. (2014). Dark tourism, thanatourism, and dissonance in heritage tourism management: New directions in contemporary tourism research. *Journal of Heritage Tourism, 9*(2), 166–182. https://doi.org/10.1080/1743873X. 2013.807266

Hartmann, R., Lennon, J., Reynolds, D., Rice, A., Rosenbaum, A., & Stone, P. (2018). The history of dark tourism. *Journal of Tourism History, 10*(3), 269–295. https://doi.org/10.1080/1755182X.2018.1545394

Hou, R. (2015). Beijing: Its characteristics of historical development and transformation. In R. Hou (Ed.), *Symposium on Chinese historical geography* (pp. 1–29). Springer.

Hsiung, J., & Levine, S. (1992). *China's bitter victory: The war with Japan, 1937–1947*. M.E. Sharpe. Reprinted by Routledge 2015.

Hung, C.-H. (2018). Communist tradition and market forces: Red tourism and politics in contemporary China. *Journal of Contemporary China, 27*(1–4), 902–923. https://doi.org/10.1080/10670564.2018.1488105

Jiunn, Y. (2011). *Jiunn's special: Lu Gou Qiao – Marco Polo Bridge*. https://jiunnsspecial.blogspot.com/2011/05/lu-gou-qiao-marco-polo-bridge/htl

Kelly. (2017). *Marco Polo Bridge (Lugou Bridge)*. https://chinahighlights.com/beijing/attraction/marco-polo-bridge-beijing-lugou-bridge/htlm

Kin, L. M. (1981). *Marco Polo in China*. Kongsway International Publications Ltd.

Latham, R. (1958). *The travels of Marco Polo*. Translated and with an introduction by Ronald Latham. Penguin Books. (Original work published 1950)

Lennon, J., & Foley, M. (2000). *Dark tourism: The attraction of death and disaster*. Continuum.

Li, Y., Hu, Z. Y., & Zhang, C. Z. (2010). Red tourism: Sustaining communist identity in a rapidly changing China. *Journal of Tourism and Cultural Change, 8*(1–2), 101–119. https://doi.org/10.1080/14766825.2010.493939

Light, D. (2017, August). Progress in dark tourism and thanatourism research: An uneasy relationship with heritage tourism. *Tourism Management, 61*, 275–301. https://doi.org/10.1016/j.tourman.2017.01.011

Mitter, R. (2013). *Forgotten Ally: China's World War II, 1937–1945*. Houghton Mifflin Harcourt.

People's Daily. (2017, August 3). The Stone Lions of Lugou Bridge: When were they carved? *Youlin Magazine*. https://www.youlinmagazine.com/article/the-stone-lions-of-lugou-bridge-when-were-they-carved/

Prideaux, B., Timothy, D., & Chon, K. (2008). *Cultural and heritage tourism in Asia and the Pacific*. Routledge.

Seaton, A. (2009). Thanatourism and its discontents: An appraisal of a decade's work with some future issues and directions. In T. Jamal & M. Robinson (Eds.), *The Sage handbook of tourism studies* (pp. 521–542). Sage.

Sharpley, R., & Stone, P. (2009). *The darker side of travel: The theory and practice of dark tourism*. Channel View.

Stone, P. (2005). Dark tourism consumption – a call for research. *e-review of Tourism Research, 3*(5), 109–117.

Stone, P. (2006). A dark tourism spectrum: Towards a typology of death and macabre related tourist sites, attractions and exhibitions. *Tourism, 54*(2), 145–160.

Stone, P. (2013). Dark tourism scholarship: A critical review. *International Journal of Culture, Tourism and Hospitality Research, 7*(3), 307–318. https://doi.org/10.1108/IJCTHR-06-2013-0039

Stone, P., Hartmann, R., Seaton, A., Sharpley, R., & White, L. (2018). *Palgrave handbook of dark tourism studies*. Palgrave Macmillan.

Stone, P., & Sharpley, R. (2008). Consuming dark tourism: A thanatological perspective. *Annals of Tourism Research, 36*(23), 574–595. https://doi.org/10.1016/j.annals.2008.02.003

Su, M. M., & Wall, G. (2012). Global–local relationships and governance issues at the Great Wall World Heritage Site, China. *Journal of Sustainable Tourism, 20*(8), 1067–1086. https://doi.org/10.1080/09669582.2012.671330

Tang, Y. (2014a). Travel motivation, destination image and visitor satisfaction of international tourist after the 2008 Wenchuan earthquake: A structural modeling approach. *Asia Pacific Journal of Tourism Research, 19*(21), 1260–1277. https://doi.org/10.1080/10941665.2013.844181

Tang, Y. (2014b). Dark touristic perception: Motivation, experience and benefits interpreted from the visit to seismic memorial sites in Sichuan Province. *Journal of Mountain Science, 11*(5), 1326–1341. https://doi.org/10.1007/s11629-013-2857-4

Tang, Y. (2018a). Dark tourism to Seismic memorial sites. In P. Stone, R. Hartmann, T. Seaton, R. Sharpley, & L. White (Ed.), *Palgrave handbook of dark tourism studies* (pp. 423–441). Palgrave Macmillan.

Tang, Y. (2018b). Contested narratives at the Hanwang earthquake memorial park: Where ghost industrial town and seismic memorial meet. *GEO Heritage*. https://doi.org/10.1007/s12371-018-0309-9

Timothy, D., & Boyd, S. (2003). *Heritage tourism*. Pearson Education.

Tunbridge, J. E., & Ashworth, G. J. (1996). *Dissonant heritage: The management of the past as a resource in conflict*. Wiley.

Wall, G., & Zhao, N. (2009, March 25). *Understanding red tourism*. Paper presented at the Annual Meeting of the Association of American Geographers (AAG), Las Vegas.

Wang, J., & Luo, X. (2018). Resident perception of dark tourism impact: The case of Beichuan County, China. *Journal of Tourism and Cultural Change*, *16*(5), 463–481. https://doi.org/10.1080/14766825.2017.1345918

Williams, S. (2009). The past as a foreign country: Heritage attractions in contemporary tourism. In *Tourism geography: A new synthesis* (2nd ed., pp. 236–257). Routledge.

Zheng, C., Zhang, J., Qian, L., Jurowski, C., Zhang, H., & Yan, B. (2018). The inner struggle of visiting 'dark tourism' sites: Examining the relationship between perceived constraints and motivations. *Current Issues in Tourism*, *21*(15), 1710–1727. https://doi.org/10.1080/13683500.2016.1220512

7 Remembering Japanese American confinement

Memorial practices at Amache and Manzanar

Whitney J. Peterson and Bonnie J. Clark

7.1 Introduction

No discussion of the memorial sites of the Pacific War would be complete without a reflection on the sites where the U.S. government incarcerated Japanese American individuals and families during World War II. In popular media, this history often begins with the U.S. reaction to the bombing of Pearl Harbor by the Japanese Army, but the incarceration is best contextualized by deep-seated racism and exclusionary U.S. policy directed at people of Asian descent in the United States, beginning with the first wave of Chinese emigration. The Chinese Exclusion Act of 1882 was the first U.S. legislation to single out a nationality for immigration restrictions. It was only subsequent to its passage that Japanese Americans began to emigrate to the United States in any number. The first-generation Japanese Americans, or *Issei*, worked in many of the same labor-intensive jobs once worked by overseas Chinese but also gravitated to agriculture. Like the Chinese, as Asians in America, they were classified as "aliens ineligible for citizenship." After 1913, such Aliens were not allowed to purchase land in California, and many other western states quickly passed similar laws.

The history of institutional racism is foundational for framing the U.S. government's response to the attack on Pearl Harbor and President Franklin Delano Roosevelt's signing of Executive Order 9066 in February 1942 (Robinson 2001). EO 9066 further facilitated anti-Asian sentiment focused on Japanese Americans in the United States and allowed the U.S. military to "exclude" anyone, including citizens, from any area of the United States without due process if deemed a military necessity. Although the text of EO 9066 makes no mention of Japanese Americans, it was generally understood that they were to be its target.

Within two weeks of the signing of the executive order, the U.S. Department of War established an exclusion zone along the West Coast for Americans of Japanese ancestry, two-thirds of whom were born in the United States and were therefore citizens. The military quickly passed over the administration of the removal and incarceration of over 120,000 Japanese Americans within the exclusion zone to a civilian agency, the War Relocation Authority, or WRA. From 1941 to 1946, Japanese Americans whom the U.S. government forcibly removed from their homes were incarcerated in various forms of government-controlled

DOI: 10.4324/9780367823795-11

confinement facilities individually and sometimes jointly managed by differing government bodies, including the War Department, the Justice Department, and the Department of Interior. The intricate and hastily built system of forced removal, incarceration, and relocation facilitated years of constant change and uncertainty for Japanese Americans and an overall dislocation of families and communities.

The arrest of leaders in Japanese American communities, often heads of households, had begun immediately following the attack on Pearl Harbor. The mass removal took place in rolling sets of exclusion orders for each census district within the exclusion zone. Once an exclusion order was issued in a particular area, the so-called evacuees typically had just one week to put their affairs in order. They were allowed to bring only what they could carry, and the decision of what to bring and what to leave behind caused significant emotional hardship and lasting economic devastation for Japanese American communities. After being forced from their homes, the majority of Japanese Americans were first confined in make-shift accommodations called by the WRA, assembly centers. They were, however, guarded detention centers, quickly constructed at makeshift sites such as fairgrounds. After several months at an assembly center, most incarcerees were taken to one of the ten quickly built WRA confinement camps in the interior of the United States where most Japanese Americans were incarcerated throughout the war. It is these sites that the majority of *Issei* and their citizen children, known as *Nissei*, experienced and later memorialized.

The WRA confinement camps follow well the trajectory proposed by Kenneth Foote (2003) for landscapes of violence and tragedy in the United States. Foote notes that such sites may cycle through four different fates: obliteration, rectification, designation, and sanctification (see also Hartmann 2014:175). The WRA camps were among the case sites chosen by Foote to demonstrate his model and the fact that no one outcome is ever final (Foote 2003:304–308).

Immediately following their closure, the WRA camps were largely physically obliterated under agency orders. Dismantling these sites removed material reminders of these locales, each of which had been sizable population centers (e.g., Heart Mountain in Wyoming was the third largest city in that state during World War II). Immediately following the war, discussion of the experience was likewise muted within the Japanese American community and the American public sphere as a whole.

During the civil rights movements of the 1960s, younger Japanese Americans joined outspoken camp survivors in pushing for rectification, not so many of the physical sites themselves but of the community. Calls came for investigation of the causes of internment and Congressional findings supported the cause for reparation. The passage of the 1988 Civil Liberties Act led to both a presidential apology and remuneration (checks of $20,000 were issued to camp survivors). While the passage led to a greater public awareness of internment, within the community the public hearings that were part of the process did even more to galvanize memorialization. They created the impetus and forum for community members to speak out about their experience.

The physical sites of Japanese American incarceration became touchstones during rectification efforts. As explored more fully below, they became sites for personal, family, and community pilgrimage. These efforts began with physically marking the sites, often through a highway sign or an on-site memorial (Foote 2003:306). Community members and some local residents began to push for the designation of these sites as official heritage sites, an effort aided by the federal recognition of the Civil Liberties Act. The first site to be recognized at the federal level was Manzanar, which became a unit of the National Park Service (NPS) in 1992. Since that time, two more of the WRA sites – Tule Lake and Minidoka – have become part of the NPS system. One additional site, the Granada Relocation Center (better known as Amache), became a National Historic Site during the writing of this chapter and is currently (2023) transitioning to National Park management. Others, such as Heart Mountain and Topaz, have been preserved through partnerships between local caretakers and the survivor and descendant community.

Once they became sites of pilgrimage, the WRA camps would seem to have moved into the final stage of Foote's model, sanctification (2003). This claim is supported by the work of religious studies scholar Jane Iwamura. The wartime incarceration of Japanese Americans was one of those crises that challenged national meaning and thus "require the formation of new myths, symbols, and rituals to support a transformed sense of a people's self-identity" (Iwamura 2007:942). Iwamura suggests that Japanese Americans have created a civil religion founded on community duty to recount the injustices of the internment and make sure it does not happen again. This civic religion is "supported by its own set of sacred texts, sites, and rituals" (2007:942). Key among those sites are the ten WRA incarceration camps.

This chapter explores memorial practices at two of these sites – Manzanar in California, and Amache in Colorado. These two case studies were chosen in part because through most of the past three decades, they represented two different models of heritage management under which many of the camps currently fall: federal oversight (Manzanar) and grassroots management (Amache). They also differ in their exposure. Manzanar has long been in the national imagination, likely due to its proximity to the large Japanese American community in Los Angeles and the popularity of publications like *Farewell to Manzanar* (Houston and Houston 1973) and *Born Free and Equal* (Adams 1944), with later depictions in popular media such as the 1999 award-winning film *Snow Falling on Cedars* based on the book by David Guterson. Amache, on the other hand, was largely unknown to even Colorado residents until quite recently. No published book-length works on the site were available until 2003 (Harvey), and that single history remained alone until 2020. Broader statewide exposure came in 2012 when Amache became a featured community at the History Colorado Center in Denver. These case studies were also chosen based on the experience of the authors. Peterson worked for the National Park Service at Manzanar and built on that experience in her thesis research on photography at both Manzanar and Amache (Peterson 2018). Since 2005, Clark has led the University of Denver research project at Amache (Clark 2019a, 2019b).

This chapter focuses on three different memorial practices critical to the heritage processes taking place at both sites – pilgrimage, the creation and curation of photo albums, and archaeology. Although different modalities, these practices exhibit a complex interweaving of tangible and intangible heritage in reclaiming a still-shadowed legacy of civic injustice.

7.2 Pilgrimage to the past

Pilgrimage is a time-honored form of memorialization woven through many religious traditions, including those most often found among *Nikkei* populations – Buddhism and Christianity. It is often focused on honoring the dead, especially martyrs (Wheeler 1999). The first pilgrimage to a WRA confinement camp took place in 1969 at Manzanar and in the decades since has become a firmly established tradition at most of the other nine camps. It is enabled by camp-specific groups, and since 2018 encouraged by a website, Japanese American Memorial Pilgrimages, devoted to documenting the pilgrimages and encouraging younger generations of Japanese Americans to attend (www.jampilgrimages.com).

A good source on pilgrimages comes from newspaper accounts that often quote participants. In reflecting on these events, some liken them to family reunion (Yamato 2017), while others note the central role of "honoring the ancestors" (Bowean 2019). These events combine festival, political forum, and religious ceremony (Iwamura 2007:938). Manzanar and Amache are among those camps with yearly pilgrimages and a comparison of the two highlights both how they differ and what they share as memorial practices. For further comparison, readers can refer to the work of Joanne Doi (2003), who has written about pilgrimages at Tule Lake.

7.2.1 Manzanar pilgrimages

Twenty-four years after Manzanar closed, a group of Los Angeles-based college students organized a bus full of people to pilgrimage to the former site of confinement. Driving north on Highway 395 (the site is just over 200 miles from L.A.), they followed the path that thousands of Japanese Americans took when they were forced to leave their homes behind in 1942. The 1969 pilgrimage encompassed a growing movement in the 1960s to recognize the injustices faced by people and minorities throughout the United States. While the pilgrimage in 1969 is most widely recognized as the first, Buddhist Reverend Shinjo Nagatomi led an annual visit to the site the decade following the closure of the camp. Buddhist Reverend Sentoku Mayeda and Christian Reverend Shiochi Wakahiro continued this tradition into the 1970s (Catton and Krahe 2018).

Sue Kunitomi Embrey was among the 250 people who attended the pilgrimage in 1969, returning to Manzanar for the first time after being incarcerated there as a young woman. Politically active in the Los Angeles area, the pilgrimage was a further catalyst in her advocacy to speak out and educate people about Japanese

American experiences of World War II incarceration. Embrey went on to form the Manzanar Committee, which became instrumental in gaining government protections for the site, culminating in the establishment of Manzanar as a National Historic Site under the management of the U.S. National Park Service in 1992. Throughout this time, the Manzanar Committee continued to sponsor the annual Manzanar and continues to do so today.

As Iwamura notes, "Rituals play an important role in the preservation of memory as they involve the active bodily participation of all those who are present" (Iwamura 2007:950). For many people, the act of returning to Manzanar is an act of remembering. For some, the experience brings back first-hand recollections of their incarceration. Robert A. Nakamura remembers his experience at the first pilgrimage in 1969.

> And so wandering around, I was able to recall, "I used to play here, we used to do that there." So very, very emotional . . . Like a lot of the *Nisei*, the whole experience was repressed, and just came out all at once at the pilgrimage.
>
> (Nakamura 2011)

The pilgrimage experience takes on a different form of remembrance for people who were not incarcerated in the camp: remembering the individuals who were forced to leave their homes and live within the one-square-mile barbed wire fence. The exposed landscape serves as a canvas for visualizing the harsh conditions, crowded confines, and isolated surroundings – a way of acknowledging past wrongs and those who endured their lasting impact.

The number of buses driving north on Highway 395 for the annual pilgrimage has grown since 1969. The main event at the pilgrimage remains centered on Manzanar's cemetery monument where 14 people are buried. Each year, people gather at the monument as flags are raised representing each of the ten camps where people were incarcerated. Survivors of the camps are asked to stand. The program changes each year as new speakers take the podium to address the crowd – often former incarcerees, activists, and artists (Iwamura 2007:938). The event includes a traditional *taiko* drum performance. Every year there is a combined Buddhist and Protestant service at the cemetery monument where people leave flower offerings, as shown in Figure 7.1 (Catton and Krahe 2018:68). Following the event, attendees are encouraged to join in another dance – *Tanko Bushi* (coal miners dance) – a traditional dance to honor the dead. Mitsue Nishio, who was confined at Manzanar as a young woman, describes the dance, which also takes place at other events such as Nisei Week, as a source of happiness and community connection (Nishio 2014).

Following the legacy of the organizers of the 1969 Manzanar pilgrimage, the Nikkei Student Unions from a number of Los Angeles–based universities co-sponsor an event that facilitates discussion. The setting provides a space of reflection and conversation for younger generations of Japanese Americans, survivors of the camps, and increasingly diverse publics who attend the pilgrimage. At the 50th anniversary of the first Manzanar Pilgrimage in 2019, Bruce Embrey

Figure 7.1 An example of the ceremony that takes place each year at Manzanar's cemetery.
Photograph by Jeffery Burton.

Source: image courtesy of the National Park Service

Co-chair of the Manzanar Committee, Sue Kunitomi Embrey's son recalls the legacy of the event.

> The Pilgrimages were a quest, searching for the truth of what happened, led mostly by young *Sansei* (third generation Japanese Americans). As it got more established, the Pilgrimage became a safe place for the survivors of camp to talk story, revealing the atrocities of camp, and educating the younger generations and broader public about our story.
>
> (Manzanar Committee 2019)

The annual Manzanar pilgrimage continues to be part of the processes of sanctification. Over time the pilgrimage has contributed to the broader convergence of various groups and stakeholder involvement at the site culminating in today's federal management as a National Historic Site. The Manzanar pilgrimage began as a grassroots movement and over time has also marked significant milestones in acquiring government-designated protections for the site. Following the 1969 pilgrimage, Manzanar Committee initiated steps to preserve the Manzanar and better facilitate the annual pilgrimage by establishing an agreement with the Los Angeles Department of Water and Power, the entity that owned the land at Manzanar (Catton and Krahe 2018:69). The 1985 pilgrimage provided a platform to officially present the site as a National Historic Landmark, and the event included representatives from the National Park Service and the City of Los Angeles who gave addresses at the ceremony. The plaque for the National Historic Landmark was installed and unveiled at the pilgrimage the following year (Catton and Krahe 2018:85). After the establishment of Manzanar as a National Historic Site on March 3, 1992, the 24th pilgrimage had the largest attendance ever recorded. The local newspaper recorded 20 busloads of people arriving from Los Angeles with an estimated 3,000 people attending the ceremony (Catton and Krahe 2018:106). Since the official establishment of Manzanar under the National Park Service, Manzanar Committee has continued to sponsor the annual pilgrimage working with the National Park Service as a supporting entity. In addition to traditional practices of honoring ancestors and survivors, the annual pilgrimage has become a space where significant milestones in the development of the site are celebrated, such as the grand opening of the visitor center at the pilgrimage in 2004.

7.2.2 Amache pilgrimages

In general, the availability of published information about the Amache pilgrimage echoes that of the two camps as a whole; Manzanar has a larger public presence. Erin Saar Hanes' master's thesis about the role of Amache in the traditions of the survivor and descendant community (2012) contains quite a few interviews that discuss pilgrimages at the site. It is a primary source for this reflection, along with (the co-author) Clark's experience of having attended the Amache pilgrimage since 2006. The first pilgrimage to Amache was 1976, with yearly pilgrimages beginning in 1983 (Hanes 2012). Manzanar is much closer to a critical mass of survivors and descendants than Amache and so its earlier pilgrimage date likely reflects that spatial association. The first Amache pilgrimage was associated with The Asian American Community Action Research Project, a civil rights activist group (Hanes 2012). A visit to the cemetery with flower offerings was integral to that first pilgrimage, as was a potluck meal afterward (Hanes 2012:139) Those two elements – honoring those buried at the Amache cemetery with offerings followed by an informal potluck meal – remained central when the yearly pilgrimages began. Like the first, they were organized by groups out of Denver, at first by the Denver Central Optimist Club, a fraternal organization made up predominantly of former Amache incarcerees. As the population of that organization began to dwindle, the group was folded into the Japanese Association of Colorado, now known as the Nikkeijin Kai of Colorado. The Nikkeijin Kai plays a critical role in organizing transportation from Denver to Amache, which is over 200 miles to the southeast of the city. Typically, at least one bus picks up attendees at two Denver-area Japanese American congregations: the Tri-State/Denver Buddhist temple and Simpson United Methodist Church.

Members of those congregations and other Japanese Americans from Colorado join local residents, especially the members of the Amache Preservation Society, as the core of the pilgrimage population. Some former Amache incarcerees are among those Colorado residents, but the majority live outside of the state, especially in California. A handful fly into Denver, often with family members, to attend the pilgrimage. The number of Amache survivors who attend swell for special anniversaries or events, such as the revealing of the National Historic Landmark plaque in 2006, and 2022, the year following Amache's designation as a National Historic Site. In recent years that core pilgrimage population has been more consistently joined by people who are not Japanese American, including the media and people committed to social justice or learning more about internment. Attendance is modest (especially compared to Manzanar), with participation for the last few years estimated by organizers to range between 100 and 200 people.

A typical Amache pilgrimage begins at the cemetery, where currently there are nine marked graves, as well as one devoted to "Evacuees Unknown." Adjacent to the graves is a brick structure which houses a granite memorial. The memorial was installed in September 1945 just before the camp closed and its role was, like that of the pilgrimages, to honor those who had passed away at the camp. The building is quite small so the pilgrimage ceremony centers on an *ireito* monument installed in 1983 by the Optimists Club. Capped with the words "Amache Remembered," it also features the names of all of the soldiers from Amache who were killed in World War II. Led by a priest from the Tri-State/Denver Buddhist Temple, the ceremony typically features prayers, incense and flower offerings, as well as a dharma talk (Figure 7.2). The ceremony honors all who were confined at the site, not just

Figure 7.2 A priest from the Tri-State/Denver Buddhist Temple leads the ceremony at the Amache cemetery during the 2014 pilgrimage.

Source: photograph by Bonnie J. Clark

those buried there. Reverend Carol O'Dowd, who has presided over several of the Amache cemetery ceremonies, noted in an interview that the ceremony is used to create sacred space (Hanes 2012), a sentiment often explained to pilgrimage attendees as the ceremony starts. By linking those who were at Amache with those in attendance, the pilgrimage ceremony creates a "relationship that transcends time" (Hanes 2012:151).

A lunch in the nearby town of Granada follows the ceremony at the cemetery. Participants are still encouraged to contribute a potluck dish, but the local residents in Granada (through the Amache Preservation Society organized under the Granada school district) now provide the lion's share of food. This is an acknowledgment that the rising number of people coming from outside the core congregations may not be aware of or able to bring food to contribute. A blessing on the food is typically given by the pastor from Simpson United Methodist, acknowledging the important role of Christian beliefs in the community. Participants sit at the long tables in the school cafeteria, and one often sees people from different demographics connect through conversation. Sometimes there is a formal program at the school, but more often participants are given an update about activities happening in the Amache community and are addressed by community leaders or politicians in attendance.

Following lunch, participants are encouraged to visit the Amache museum, located in Granada or to go back to visit the site if they have their own transportation. If participants have a personal or familial Amache connection and would like to visit their barracks, the archaeology crew or Granada residents try to accommodate those requests. Returning to these locales as family groupings has long been an important part of the pilgrimages, but few can do so without some help, as the individual blocks are unmarked and look strikingly similar. This becomes even more true as families return after a long absence or without their survivor elders. Occasionally a site tour is incorporated into the pilgrimage, typically when developments have taken place on site, such as when historic structures have been returned to camp or replica structures built.

The bus from Denver must return mid-afternoon, so transportation restricts the time allotted for many of the formal pilgrimage activities. With more families coming from out of state, an attempt has been made to expand the pilgrimage to a full day, with an evening dinner for those who have opted to stay overnight near Amache. Typically, this dinner is organized both by the Amache Preservation Society and by the Amache Alliance, an umbrella group with representatives from many of the Amache-related grassroots groups. With no one organization organizing the pilgrimage, these efforts continue to evolve.

7.2.3 Synthesis

One aspect both Amache and Manzanar share, along with only one other camp – Rohwer in Arkansas – is an extant cemetery. Given the importance of the physical remains of past individuals to the concept of pilgrimage, it is unsurprising that these pilgrimages are anchored in these locales. This makes pilgrimages at both sites different from those at the WRA sites that lack cemeteries. Although participants at the other sites honor the ancestors, they must do so in another location, one that has perhaps less resonance in the traditional sense of pilgrimage.

Interviews in Hanes' thesis echo what has been written about pilgrimage at other camps. One key aspect of the experience is the physical experience of place. For example, a woman whose father was confined at Amache notes that "part of memorializing and remembrance is witnessing the location. It is important for them to see what it was like for the former internees by seeing and experiencing the physicality of the site" (Hanes 2012:137). Another critical element of the pilgrimages is their intergenerational nature. Minoru Tonai, who has led the most active group of former Amache incarcerees – the Amache Historical Society – noted in an interview that "pilgrimages have created a bridge between the older and younger generations of the community, with many younger members of the community learning about and experiencing Amache for the first time" (Hanes 2012:147).

7.3 Photo albums as memorial practices

Despite significant limitations in photographing their lives in the camps during World War II, many Japanese Americans who were incarcerated have photographs from the time. Each year, the museums at Manzanar and Amache receive donations from former incarcerees and their descendants. These collections increasingly include photographs and photo albums. The mnemonic role of photographs and the camera as a memory device has seeded fertile ground for discourse and contention surrounding some of the most fundamental factors in negotiating and remembering the incarceration of Japanese Americans during World War II. The photographic landscape of this history raises questions about power, perspective, and agency, illustrative of the core injustices that made the incarceration possible. The material presented in this section of the chapter derives in large part from co-author Peterson's thesis research (2018).

While a significant amount of scholarly attention has been focused on the government-produced photographs of this time, photo albums that Japanese Americans created documenting their lives in confinement have received less critical attention. While many studies focus on exploring the visual representation of the incarceration in the images, few have observed how the photographs themselves are part of commemorative practices. Through their creation, exchange, and preservation, the photo albums are part of how Japanese Americans have negotiated the memory and legacy of this history with their families, communities, and the broader public. As these albums and their photographs have made their way back to the sites where they were created at museums located at former sites of confinement, they provide a framework for exploring how these places are part of ongoing processes of commemoration and memory.

The focus on the visual power of images has influenced the often taken-for-granted material nature of photographs. Visual Anthropologist Elizabeth Edwards' approach to the study of images as objects provides an opportunity to understand the ways in which photographs are situated in the cultural processes of negotiating community identities. Through her study of photographs as material objects, she explores how they are part of acts of reclaiming histories and narratives originally created for hegemonic colonial or scientific purposes. Quoting Brown and Peers, she writes that community members utilize photographs to "articulate to themselves their experiences of the past and, ultimately, to speak to their children about the strength of their community" (Edwards 2011:181).

Shifting away from an analysis of what photographs visually signify and instead observing them as material objects reveals how photographs are part of complicated social networks and practices of memorialization. Quoting Brent Plate, Iwamura notes, "Memory is not an activity of the mind only, but of the body and of the minds and bodies of others . . . Remembering is a dynamic, interactive process" (Iwamura 950). Rather than a static form of documentation and information transfer, photo albums are part of ongoing acts of interpretation and performance and larger processes of memorialization. In Kristen Emiko McAllister's *Terrain of Memory: A Japanese Canadian Memorial Project*, McAllister writes,

> Rituals and social practices – whether funerals, commemorations of the war dead, or looking through family photograph albums – affirm a shared origin; they gather us together to affirm our communal ties. These events, practices, and institutions selectively identify historical figures and events that shape our collective identities, symbolize the values and goals we share, and form the basis for imagining and planning for a future together.
>
> (12)

The context of contradiction and hardship that Japanese Americans faced in photographically documenting their lives during World War II has shaped the ways that people have engaged in processes of commemoration through photographs. While the federal government at times prohibited Japanese Americans from owning cameras and documenting their lives during World War II, they simultaneously hired professional photographers to meticulously document the incarceration (Figure 7.3). These images were used not only for documentation but also for propaganda. Writing on the back of many of these original images found in government archives indicates that certain photographs and accompanying captions were either altered or "impounded," while other images that aligned with government messaging were shared with the public. Throughout the war Japanese American photographers such as Toyo Miyatake at Manzanar and Jack Muro at Amache pushed back on the U.S. government's prohibition of photography in the camps. They secretly fashioned cameras and dark rooms and documented the lives of their new confined communities behind barbed wire.

Throughout the incarceration, Japanese Americans who like many Americans embraced "Kodak culture" and had been accustomed to photographing their lives, looked for ways to continue this practice. Following the Japanese attack on Pearl Harbor on December 7, 1941, the U.S. government seized items from Japanese Americans who were non-citizens, including radios and cameras. As Japanese Americans on the West Coast were forced to leave their homes for confinement

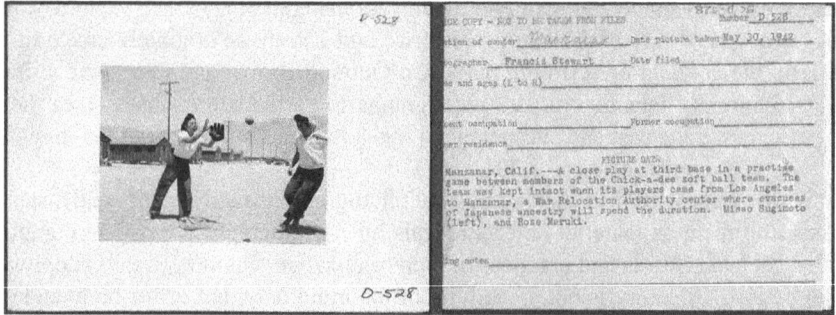

Figure 7.3 Front and back of a photograph taken by government photographer Francis Stewart of Rosie Maruki Kakkuchi playing softball at Manzanar (WRA no. D-528, War Relocation Authority Photographs of Japanese-American Evacuation and Resettlement Collection, The Bancroft Library, University of California, Berkeley)

in the camps, cameras were added to the list of contraband and not allowed to be brought into the camps. While regulations changed over the course of the war, people's ability to document and photograph their lives was limited. A letter published in Amache's camp newspaper expressed the frustration of a woman, Mrs. T Muta, attempting to retrieve her camera in the custody of officials in Los Angeles writing, "The cameras are of no value to the owner if left in Los Angeles, and to buy another one is impossible" (Granada Pioneer December 19, 1942).

Photo studios were eventually established in each of the ten camps, but supplies and access remained limited (Figure 7.4). For those confined at Manzanar, Toyo Miyatake and his son Archie Miyatake were well known and provided many people with photographs while they were unable to take their own. Government records at Amache indicate a significant demand for photographs, even those produced by the U.S. government stating that the Amache Co-op board "appears to have been shortsighted in refusing to carry out a function of distributing photographs earnestly desired by many residents" (Rademaker 1943:3). An article in the Amache newspaper announced the sale of the photographs: "An assortment of some 500 official WRA photographs, illustrating Amache camp life, is expected to go on sale at the local Co-op store the latter part of this month" (Granada Pioneer February 2, 1944). People eventually managed to acquire photographs and sometimes cameras. The challenges and ingenuity involved in taking and acquiring photographs provide context for the ways people have engaged with photo albums over time and their role in remembering this history.

Interviews with people who created or donated the albums to museums at former sites of confinement shed light on the ways these albums are part of remembering this history with people's families, communities, and the public. In the 75 years since the incarceration, people's photo albums have gone through different stages of use and transformation, sometimes remaining unused in storage, as well as both social and solitary acts of reflection and family community discourse.

Figure 7.4 Photograph from Hatsume Akaki's photo album of workers at the Co-operative Enterprises at Amache.

Source: image courtesy of the Amache Preservation Society

Photographic documentation of people's lives in confinement persisted despite silences surrounding the incarceration in the decades following World War II. As individuals and community organizations became more vocal about this history and pushed to preserve sites associated with the incarceration, photo albums have increasingly become involved in more public-facing discourse and processes of commemoration.

Many photo albums donated to the Amache Museum and Manzanar National Historic Site have names written on the backs or front indicating that acquiring the photographs involved exchanging the photographs with other people (Figures 7.5 and 7.6). In an interview with Tami Kasamatsu, who was incarcerated at Manzanar as a young woman, she mentioned that she acquired many of her photographs from other people. "Well, whenever anybody takes pictures, I think somebody had their own cameras too, and so whenever they took the pictures, I asked for copies" (Peterson 2018:37).

These processes of exchange continued after the war. Rosie Kakuuchi, who was also incarcerated at Manzanar, recalls getting together with a group of women who had formed a social group while confined at Manzanar after the war. When they got together, they would often exchange photographs from camp with one another. For people who were not the original creators of the donated photo albums, acts of recreating and donating the albums were also involved in these processes of connecting with family and camp communities. Joyce Seippel, who was confined at Manzanar as a young girl, came across her cousin Katusmi Taniguchi's photo album in his home after he passed away. With the negatives that she found, she created a digital album and physical book of the photographs as shown

Figure 7.5 Front and back of photograph in Hatsumi Akaki's photo album donated to the Amache Museum by her nephew, Robert Akaki.

Source: image courtesy of the Amache Preservation Society

Figure 7.6 Page from Rose Tanaka's photo album that she donated to Manzanar National Historic Site. Writing is visible on the front and backs of photographs.

Source: photograph by Whitney J. Peterson from Manzanar National Historic Site, Rose Tanaka Collection

in Figure 7.7, working with her cousin, Katsumi's wife, to identify people in the photographs.

For many Japanese Americans connected to World War II incarceration, rituals surrounding the memorialization of the sites and experiences are not necessarily involved in recalling their own experiences but a way of uncovering unknown stories or discourse silenced during the decades following World War II. As Joyce Seippel went through the photographs of her cousin by marriage whom she had not known while she was confined at Manzanar, she came across a photograph of her

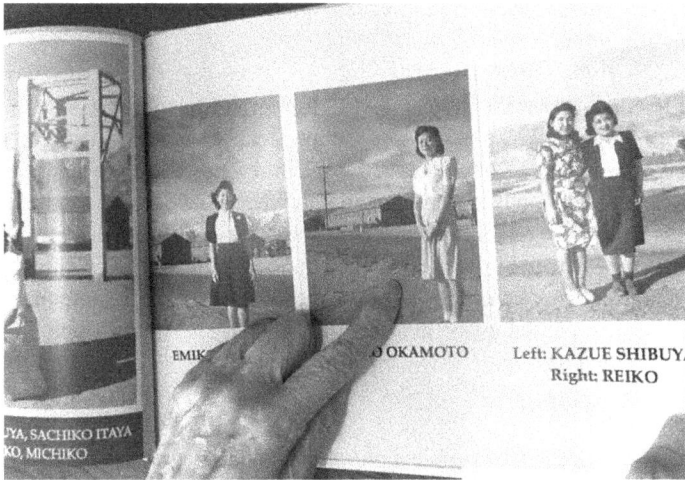

Figure 7.7 Photo book that Joyce Seippel put together with her cousin Katsumi Taniguchi's negatives.

Source: photograph by Whitney J. Peterson

mother. Similarly, as Rosie Kakuuchi has been reorganizing her family's albums including their Manzanar experience, she came across other photographs that filled gaps in her knowledge of her family's experience. While attending a family funeral she discovered a photograph of her sister who had died during childbirth while confined at Manzanar. This photograph provided insight into part of her sister's life she had not known and felt it was important to include in her sister's portion of the album. Rose Tanaka, confined at Manzanar as a young woman, describes how looking at her photographs with her children and grandchildren and visiting the Manzanar site with them has provided an opportunity to share with them something she believes is important for them to know about. She states, "My kids they come, and they like to go through and browse through the books that I have here. And so, anyhow, I think they might profit from knowledge about the background, and they are all good kids" (Peterson 2018:195).

As the albums have been part of practices in exploring family and community knowledge, their donation, display, and preservation in the more public spaces of museums are a means to continue these social processes of discovery and memorialization through reaching broader communities of people connected to this history. Bob Akaki donated his aunt's photo album to the Amache Museum with the hope that other people would gain from seeing the photographs. Similarly, Joyce Seippel donated her cousin's photographs to Manzanar hoping that other people may recognize people in the images, perhaps similar to her experience discovering her mother in her cousin's photographs.

Alisa Lynch, Chief of Interpretation at Manzanar, has observed the power of photographs when people visit the site. During the grand opening of the Manzanar

Visitor Center, she describes a family's encounter with a photograph incorporated into the new exhibits:

> When his daughter and wife came with him grand opening day and he saw that, well, first of all . . . the daughter sobbed, cause she had never seen a picture of her father as a young man. And so sometimes there is a really deeply emotional connection that is exponential to what most of us encounter in our lives, because here is a situation where somebody has no pictures, because the government forced them from their home. They lost everything. They were sent to Japan. . . . And it's a simple picture, but it brings someone to life. And he's since passed away, but um, I think there are interactions like that, that are pretty amazing.
>
> (Peterson 2018:200)

Observing the collection of people's photographs donated to the museum has illuminated how important the ritual of photographic documentation was during the incarceration. Government-produced photographs and duplicate copies of images taken by camp photographers, such as Toyo Miyatake, demonstrate people's creativity and agency in telling their own story. The photo albums contain photographs born in an environment created by racism, contradiction, and hardship but also one of resilience, agency, and community. As people's albums have returned to the sites now established to remember these histories, they are increasingly part of more public discourse and new opportunities for memorialization, remembering, and connection.

7.4 Archaeology as memorialization

With its popular association to the deep past, archaeology might seem like an unlikely form of memorialization. Yet any archaeological study is embedded in the social context and undertaken by living people, especially what has come to be known as contemporary archaeology, studies of sites either actively occupied or within living memory. All archaeological research links past and present in a way that "orients people in their cultural experiences" (Shanks and McGuire 1996:82). It does so by revealing the details of everyday life in the past through the material remains of practices as quotidian as a meal, a play space, or a pathway worn by repeated travel. It also calls attention to top-down control revealed through city planning or administrative structures. Strongly rooted in places, archaeology can play an important role in community building and the negotiation of memory.

Since the 2000s, a growing body of archaeological studies has explored the tangible remains of 20th-century wars, including military facilities (Moshenska 2010), prisoner of war camps (Mytum and Carr 2013), and defensive structures such as bomb shelters (Archaeological Institute of America 2011). The archaeology of Japanese American confinement camps combines a concern with war-time incarceration and the impact of war on the homefront. It also fits neatly into the model hypothesized by Gabriel Moshenska for how sites of conflict intertwine with memory work (2010). Drawing from a wide range of studies of memory dynamics, Moshenka notes that publicly engaged archaeological investigations become arenas of memory, where tangible remains combine powerfully with personal or communal memories. Once formed, these narratives of memory can work their way into more mainstream historical narratives through efforts by both stakeholders and the archaeologists or agencies they represent. In those instances, archaeology becomes not just an arena of memory but also an agency of articulation to larger memory discourses.

Moshenka goes on to propose that to be effective as both arenas and agencies for the articulation of memory, such archaeological projects require a commitment to be "consciously and proactively open and accessible to the public" (2010:45). Although quite different in their administration, the archaeology programs at both Manzanar and Amache share this commitment. The community archaeology program at Manzanar has been led by the National Park Service as part of their preservation and education programs at the site. Begun in 2003, the program was recognized in 2019 with the California Governor's Historic Preservation award. As the announcement states, "volunteers are discovering, documenting, and restoring landscape features that tell the stories of Manzanar to the site's visitors, who number over 100,000 per year" (California State Parks Office of Historic Preservation 2019). The first professional archaeology intervention at Amache was a site survey in 2003 (Carrillo and Killam 2004). Those findings laid the groundwork for community archaeology at Amache, launched in 2008 with an archaeology field school through the University of Denver (DU). The DU Amache Project continues (as of this writing in 2023), centered on a bi-annual field school. Although the DU Amache project has been more focused on post-secondary students, both programs incorporate a wide population into the archaeology fieldwork, including camp survivors, their descendants, the descendants of camp employees, local residents, and educators (Sahara 2009; Clark 2019b).

Collaborative by nature, archaeology requires teams of people who constantly negotiate basic questions like: What is this thing we have found? How was it used and by whom? In teasing out those basic questions, the significance of finds is not far off. In those instances, the knowledge of cultural insiders becomes a vital data source, with stories that find their way first into informal discussion and then often more formal interpretations. Whether it is the family member who uncovers features built by their ancestors (Alvarado 2011) or a find that helps recall a memory from camp (Clark 2019a), the recursive nature of the archaeological enterprise is fertile ground for memory work. As predicted by Moshenska's model (2010) these practices are why archaeology can be a powerful arena for memory articulation. Collaborations can move individual or family memories up the scale into communal memory or even the authorized heritage discourses of technical reports and site signage.

An example from Amache can illustrate this process. One of the many clever incarceree modifications to the camp's facilities was placing the bathhouse hot water heater on a large concrete block in order to improve water pressure in the showers. As opposed to most of the architecture of the site, these heater blocks were built by the incarcerees themselves. These are the kinds of situations at both Manzanar (Casella 2007) and Amache, where we find that incarcerees literally left their mark on the site. In Block 9L at Amache, four signatures were incised into the wet concrete of the heater block. Crews during the 2008 field season became quite familiar with this feature, as they walked by it daily on the way to where they were conducting test excavations. In the middle of the field season, the crew of undergraduate and graduate students was joined by Gary Ono and his grandson Dante Hilton-Ono. Gary had been a young boy at Amache and his family lived on the opposite side of the camp. When the heater block was pointed out to Gary, he immediately identified the bottom signature as that of his own father, Sam Ono. In 2003, the first archaeologists to study the site had documented this feature, but they were not able to discern that final signature because it is partially eroded (Figure 7.8). Gary identified it right away, primarily through the very distinctive "S" his father employed in his signature.

In reflecting on this moment, Gary notes he was "awe-struck by this chance discovery" (Ono 2009:np). He had no idea that his father had been involved in construction work at Amache, so the feature added one more layer to his family history. Later Gary would go on to research the other men whose signatures were on that block, showing that like his own they lived all across the site (Ono 2009). This find is just one of many examples of how the archaeological research at these sites illuminates many ways that incarcerees worked collectively to improve conditions not just in their own blocks, but across the site (e.g. Starke 2015). The knowledge gained about this visible reminder of Amache's incarcerees is now enshrined in the official interpretation of the site, through a way-side sign (Figure 7.9). This

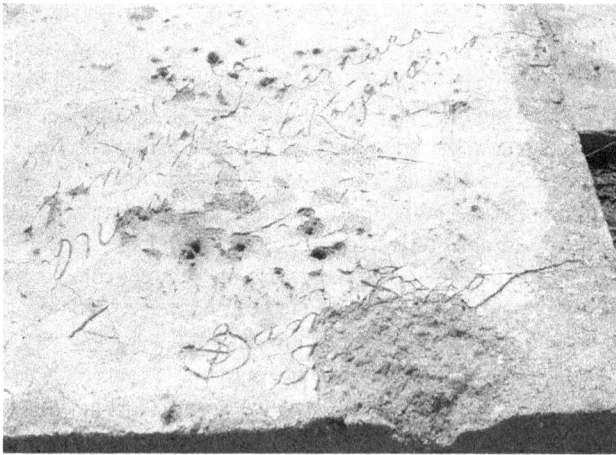

Figure 7.8 Concrete block at Amache with Sam Ono's distinctive signature near the bottom.
Source: photograph by Bonnie J. Clark

Figure 7.9 Gary's grandson Salvador Valdez-Ono poses with the sign that includes the story of his great-grandfather's signature and his grandfather's discovery of it.

Source: photograph by Bonnie J. Clark

unfolding history exemplifies the braided path from archaeological find, to family history, to interpretation, to memorial practice that community-engaged archaeology at these sites is making possible.

Reflections by community members who have chosen to take part in these archaeology projects reveal motivations for engagement that are firmly rooted in archaeology as a heritage process (Clark 2019a). For some, archaeology is an element of a longer project of recrafting a history repressed within their own families. For others, engagement in archaeology was the first step in "digging for their roots" (Carlene Tanigoshi Tinker, personal communication). As one might expect, the data uncovered in these sites helps to fill in some of the gaps where stories were not told. Dennis Fujita, who was born at Amache, wrote of his experience as a volunteer: "cataloguing of these objects has the potential to renew memories of other former internees about how they dealt with daily events in the internment camp and how these experiences shaped their lives" (2018:558–559). Yet interaction with these sites of personal or family incarceration moves beyond just data recovery into emotional engagement. Fujita writes that especially for former incarcerees, engagement in archaeology "helps us personally heal from the psychological wounds of incarceration" (2018:560). For descendants, archaeology can be a form of pilgrimage (Sahara 2009), with the same connotations of honoring the ancestors.

At these Japanese American confinement sites, archaeology can be a form of truth-telling, of drawing attention to this shadowed history. That goal is underlain by a bedrock commitment to civic justice, to learning from the mistakes of the past. As with participation in pilgrimages, this appears to be a prime motivator for many involved in archaeology. As expressed by a descendant who served as one of the Amache project's high school interns, "My heritage influenced the study of my grandfather's wartime home, and this research reinforced my commitment to the goal of justice for all" (Eijima 2018). Her older sister, also an intern with the project, expressed a similar sentiment, "Like my grandparents and many other former internees, I feel obligated to fight injustice and uphold the Constitution" (Eijima 2016).

7.5 Conclusion

As Foote notes, "Heritage serves as a symbolic foundation of collective identity. It answers the question 'who we are' by focusing on 'where we came from' and on the places and sites that inscribe this vision of historical identity on the landscape" (Foote 2003:37). Practices of commemoration at former sites of confinement have contributed to and are shaped by ongoing efforts to reconcile and learn from the past and honor community accomplishments and sacrifices (Foote 2003:38). Through pilgrimages to the sites, the creation and preservation of camp photo albums, and public archaeology programs, both Amache and Manzanar foster ongoing practices that align with the Foote's status of sanctification. Historic photographs, archaeological material culture, and the cultural landscape at Amache and Manzanar are central to these practices. They are tangible mediators – part of practices of recognition and remembrance of the U.S. government's World War II incarceration of Japanese Americans.

For decades, Amache and Manzanar differed in management, yet both sites demonstrate historic and ongoing practices that align with Foote's heritage site continuum. These places also demonstrate how these states are not static. Individual and community-based practices of remembering the Japanese Americans of World War II incarceration at these sites have taken place over decades, ignited by grassroots groups, while simultaneously obstructed or silenced by others. Amache's shift, during the course of writing this chapter, from grassroots site to National Park, is an example of how memory work can change how and by whom such sites are managed.

With the increasing role of digitization and virtual means of connection, commemorative practices associated with these sites continue to transform. Digitization of museum objects, archives, site artifacts, and the digital documentation of the sites themselves have facilitated new forms of engagement and commemoration with these places and their stories. The museums at both Amache and Manzanar have ongoing digitization projects making artifacts, photographs, and oral histories available to the public. Both the physical sites have been digitally documented using laser scanning or drone photography technologies. Online repositories such as those available through Densho (www.densho.org), which collects digital records of incarceration associated with all the camps, provide resources for people to access.

Digital forms of engagement with these sites increase as they become more accessible but also as the site and communities look for ways to connect and remember when coming together at former sites of confinement or meeting with community members presents challenges. With the onslaught of the COVID-19 virus in 2020, pilgrimages to the former sites of confinement were canceled, and

Figure 7.10 Four generations at Amache. A composite photograph made from a digitized historical image and a photograph taken during the 2016 DU Amache Field School Community Open House.

Source: image courtesy of Kirsten Leong

people from around the world have engaged in virtual forms of pilgrimage programs organized by groups and institutions such as the Japanese Americans Museum of San Jose and the organization Japanese American Memorial Pilgrimages. These digital artifacts and new forms of virtual engagement raise questions about the future of commemorative practices associated with sites of Japanese Americans' World War II incarceration. How will digital processes influence practices of commemoration such as those outlined by Foote? How do we understand the role of digital commemoration as these practices become increasingly removed from the physical sites themselves?

Commemorative practices at these sites reflect the diversity of people tied to these places. Amache and Manzanar are significant not only to former incarcerees but also to their family members and intergenerational communities whose connection to these places facilitates the negotiation of identity in the world today (Figure 7.10). These places are also part of a broader consciousness beyond Japanese American communities. As central touchstones for remembering and examining civic injustice in the United States, they continue to facilitate connection to the past through the lens of present-day negotiations of what it means to be an American.

Acknowledgments

Both authors owe an incredible debt to those who have shared with us their experiences at these sites of difficult heritage. Thank you for your trust in us. Thank you to Tami Kasamatsu, Bob Akaki, Rose Tanaka, Rosie Kakuuchi, and Joyce Seippel in particular for sharing your personal stories and albums. Thank you to Jeffery Burton, the archaeologist at Manzanar, for sharing resources and images with us. Archaeological research and community engagement at Amache have been financially supported by both History Colorado's State Historical Fund, the University of Denver, Center for Community Engagement in Support of Student Learning, the Nikkejin Kai, and generous individual donors.

References

Adams, Ansel. 1944. *Born Free and Equal: The Story of Loyal Japanese-Americans.* New York: U.S. Camera.

Alvarado, Patrick. 2011. "Nishi Family Returns to Manzanar to Help Rebuild Historic Bridge at Merritt Park." *Discover Nikkei*, January 1.

Archaeological Institute of America. 2011. "Archaeology of World War II." *Archaeology* 46 (3): online unpaginated. https://archive.archaeology.org/1105/features/world_war_II_wwII_archaeology.html.

Bowean, Lolly. 2019. "What Their Grandparents Didn't Talk About: New Generation of Japanese-Americans Traces Family History in U.S.-run Internment Camps." *Chicago Tribune*, May 2.

California State Parks Office of Historic Preservation. 2019. *2019 Governor's Historic Preservation Awards: Manzanar Community Archeology Program.* California State Parks Office of Historic Preservation.

Carrillo, Richard F., and David Killam. 2004. *Camp Amache (5PW48): A Class III Intensive Field Survey of the Granada Relocation Center, Prowers County, Colorado.* Prepared by RMC Consultants, Inc. for the Town of Granada.

Casella, Eleanor Conlin. 2007. *The Archaeology of Institutional Confinement, The American Experience in Archaeological Perspective.* Gainesville: The University Press of Florida.

Catton, Theodore and Diane L. Krahe. 2018. *The Sands of Manzanar: Japanese American Confinement, Public Memory, and the National Park Service.* U.S. Department of Interior. http://npshistory.com/publications/manz/adhi.pdf

Clark, Bonnie J. 2019a. "Collaborative Archaeology as Heritage Process." *Archaeologies* 15 (3):466–480. https://doi.org/10.1007/s11759-019-09375-6.

Clark, Bonnie J. 2019b. "Making Heritage Happen: The University of Denver Amache Field School." In *Archaeologists and the Pedagogy of Heritage, Volume 1: History and Approaches*, edited by Phyllis Mauch Messenger and Susan J. Bender, 168–180. Gainesville: University Press of Florida.

Doi, Joanne. 2003. "Tule Lake Pilgrimage: Dissonant Memories, Sacred Journey." In *Revealing the Sacred in Asian and Pacific America*, edited by Jane Iwamura and Paul Spickard, 273–289. London: Routledge.

Edwards, Elizabeth. 2011. "Tracing Photography." In *Made to Be See: Perspectives on the History of Visual Anthropology*, edited by Banks, Marcus and Jay, Ruby. Chicago: University of Chicago Press.

Eijima, Riki. 2016. "Uncovering the Holes of Our Past." *Pacific Citizen*, 4 (8). https://gallery.mailchimp.com/d92895e1358a8c3a824a00b8c/files/31d15b70-d605-4015-b4ff-8a42ecd2e3ae/Oct._26_Nov._8_2018WebColor.pdf.

Eijima, Tomi. 2018. "Community, Collaboration, and Justice For All." *Pacific Citizen*, 4 (8). https://gallery.mailchimp.com/d92895e1358a8c3a824a00b8c/files/31d15b70-d605-4015-b4ff-8a42ecd2e3ae/Oct._26_Nov._8_2018WebColor.pdf.

Foote, Kenneth E. 2003. *Shadowed Ground: America's Landscapes of Violence and Tragedy.* Revised edition. Austin: University of Texas Press.

Fujita, Dennis K. 2018. "Returning to Amache: Former Japanese American Internees Assist Archaeological Research Team." *Historical Archaeology* 52 (3):553–560. https://doi.org/10.1007/s41636-018-0129-4.

Hanes, Erin M. 2012. "Opening Pandora's Box: A Traditional Cultural Property Evaluation of the Amache World War II Japanese Internment Camp, Granada, Colorado." Master of Arts, Anthropology, Sonoma State University. https://sonoma-dspace.calstate.edu/handle/10211.1/1612

Hartmann, Rudi. 2014. "Dark Tourism, Thanatourism, and Dissonance in Heritage Tourism Management: New Directions in Contemporary Tourism Research." *Journal of Heritage Tourism* 9 (2):166–182.

Harvey, Robert. 2003. *Amache: The Story of Japanese Internment in Colorado During World War II*. Dallas: Taylor Trade Publishing.

Houston, Jeanne Wakatsuki, and James D. Houston. 1973. *Farewell to Manzanar: A True Story of Japanese American Experience During and After the World War II Internment*. Boston: Houghton Mifflin.

Iwamura, Jane Naomi. 2007. "Critical Faith: Japanese Americans and the Birth of a New Civil Religion." *American Quarterly* 59 (3):937–968.

Manzanar Committee. 2019. "50th Annual Manzanar Pilgrimage/2019 Manzanar at Dusk Set for April 27, 2019." https://manzanarcommittee.org/2019/01/28/50th-pr1/.

Moshenska, Gabriel. 2010. "Working with Memory in the Archaeology of Modern Conflict." *Cambridge Archaeological Journal* 20 (1):33–48.

Mytum, Harold, and Gilly Carr, eds. 2013. *Prisoners of War: Archaeology, Memory, and Heritage of 19th- and 20th-Century Mass Internment*. New York: Springer Science+Business Media.

Nakamura, Robert A. 2011. "Robert A. Nakamura Interview." Densho Digital Archive: Friends of Manzanar Collection. https://ddr.densho.org/media/ddr-densho-1003/ddr-densho-1003-4-transcript-b15dc0bc1a.htm.

Nishio, Mitsue. 2014. "Mitsue Nishio Interview." Densho Digital Archive: Manzanar National Historic Site Collection. https://ddr.densho.org/media/ddr-manz-1/ddr-manz-1-152-1-transcript-069da48e42.htm.

Ono, Gary T. 2009. "Significant Signatures." *Discover Nikkei*, April 24. http://www.discovernikkei.org/en/journal/2009/4/24/significant-signatures/.

Peterson, Whitney. 2018. "Snapshots of Confinement: Memory and Materiality of Japanese Americans World War II-era Photo Albums." Master of Arts, Anthropology, University of Denver. https://digitalcommons.du.edu/cgi/viewcontent.cgi?article=2518&context=etd

Rademaker, John. 1943. "Granada Project Community Analysis Section." War Relocation Authority. Granada War Relocation Center.

Robinson, Greg. 2001. *By Order of the President: FDR and the Internment of Japanese Americans*. Cambridge: Harvard University Press.

Sahara, Kanji. 2009. "Pick and Shovel Pilgrimage." *Rafu Shimpo*, September 26, 3.

Shanks, Michael, and Randall H. McGuire. 1996. "The Craft of Archaeology." *American Antiquity* 61 (1):75–88.

Starke, Zachary. 2015. "Wrestling with Tradition: Japanese Activities at Amache, a World War II Incarceration Facility." Master of Arts, Anthropology, University of Denver. https://digitalcommons.du.edu/etd/1051/.

Wheeler, Bonnie. 1999. "Models of Pilgrimage: From Communitas to Confluence." *Journal of Ritual Studies* 13 (2):26–41.

Yamato, Sharon. 2017. "Through the Fire: Pilgrimage Fever." *Rafu Shimpo*, September 8, Unpaginated.

8 Disconnection and continuity of war memories and battleship Yamato

Kyungjae Jang

8.1 Introduction

In the global boom of war memories (Winter, 2001), museums, monuments, and practices are being created in Asian countries to preserve, inherit, and commemorate war memories (Frost, Vickers and Schuhmacher, 2019). The boom in memory of war in Asia is caused by economic growth, education, and increased leisure travel, and its purpose is to deliver messages about local/national building, war/peace, and the foundation for moving forward from the past.

Unlike other East Asian countries, which mainly remember traumatic events, Japan remembers the war as both victim and offender and considers how both war and peace play out at the local and global levels. This is represented in the sites of war museums and peace museums scattered throughout Japan. Many studies have focused on the similarities and differences between the two types of spaces. Allen and Sakamoto (2013) analysed the collective memory formation seen in national/small-scale facilities, centring on war museums, especially how war museums present exhibits related to the subject of peace. Through an analysis of the Peace Museum, Yamane (2017) noted the need for more messages promoting peace, and peace education. Meanwhile, Lee (2018) emphasised the necessity of contributing to the preservation of war memories and the transmission of peace messages throughout East Asia by contrasting the differences between the Yūshūkanin the Yasukuni Shrine in Tokyo and the Hiroshima Peace Memorial Hall.

However, as Allen and Sakamoto (2013) argued, the memory of war cannot be expressed as a simple dichotomy between war and peace; it is a more diverse and complex phenomenon. This is because war memories are constantly reinterpreted and reproduced, not only through the description and conceptualisation of the fixed past but also through present people's memories, forgetfulness, and decisions to exclude certain aspects of the past. The inheritance of the atomic bomb memory in Japan is a symbolic example. As Dower (2012) suggested, memories of the atomic bomb created a message of peace and fear of war in Japan, while emphasising the necessity of rebuilding the country through technology. This complexity is particularly remarkable in the regions of the country where war most drastically influenced the developments of industry, technology, and knowledge related to war. A typical place is Kure, a naval port city in Hiroshima Prefecture. It is located right

DOI: 10.4324/9780367823795-12

next to the city of Hiroshima, where the atomic bomb was dropped, but there is a big difference in the ways in which Hiroshima and Kure have inherited war memories. While Hiroshima has lived the post-war period through cut-off of the war on the occasion of one "reset" of the atomic bomb, Kure has made a post-war history by continuing the legacy of the technology and navy created for the war while attempting severance. The dual nature of the preservation of war memories at the local level is evident from the ongoing pride for war-related technology alongside the expressions of anti-war and peace values.

The Maritime History and Science Museum (also known as the Yamato Museum), which opened in Kure in 2005, is a space that symbolises Kure's complex war memories. It was created with the intent to preserve the maritime history of the region, but it also expresses how the disconnection and continuation of war memories have played out in the region, as well as how the regional identity was created and continues to the present.

This chapter looks at the complex process of how the Yamato Museum preserves war memories. First, the chapter explains the process of disconnection and continuity of war memories in Kure, and then it considers how these are reflected through the process of opening of the museum to the public and exhibition of the Yamato Museum and the Japan Maritime Self-Defence Force (JMSDF) Kure Museum. Finally, the chapter discusses the functions of the battleship Yamato and the Yamato Museum, represented through popular culture in the memory of war that has been de-contextualised and re-contextualised since the 2000s.

8.2 Disconnection and continuity of war memories in Kure

The atomic bomb caused numerous casualties and reset Hiroshima, once a war archipelago, by breaking its memory as a military city and re-establishing itself as a peace city (Yoneyama, 1999). In other words, Hiroshima's post-war period was a process of disconnection from the war, especially regarding its military memory (Uesugi, 2012).

Meanwhile, next to Hiroshima city is the military port city Kure, which belongs to the same Hiroshima prefecture, but the two cities' inheritance of war memories is largely different. While Hiroshima underwent a post-war disconnection, Kure, while attempting its own disconnect, made history by inheriting crafts and naval heritage created for war. Uesugi (2012, 2014) addressed this as well as Kure's city image created by its post-war history in terms of disconnection and continuity. Unlike Hiroshima, Kure created a contradictory post-war memory, insisting on peace (disconnection), but continuing to maintain the ability to threaten tranquillity.

Kure was unable to create a post-war memory while simultaneously embracing the contradictions of disconnection and continuity, because it was the representative naval port and war ship-building industry, and both still function as central axes. After the war's end, Kure, like Hiroshima, attempted to eliminate military memories. The Former Military Port Transition Act was enacted in 1950 in an attempt to convert military sites and industries in Kure, Yokosuka, Sasebo, and

Maizuru, which were the headquarters of wartime navy centres. The Act was officially implemented in Kure on the 28th of June 1950, with 95 per cent of its residents in favour. However, three days before the law's enforcement, the Korean War broke out and Kure was forced to once more serve as a military city.

Later, when the JMSDF was founded, the Kure District Headquarters opened in 1954, and the Self-Defence Forces once again used the major naval facilities and land that were supposed to be converted. That same year, Hiroshima's newspaper lamented the failure of the disconnection: 'Kure is now on the path of "military port"' (Uesugi, 2014, p. 126). In the industry, heavy naval enterprises, such as shipyards, entered the former naval arsenal area, and as a result, Kure's urbanity was maintained as a space centred on the continuation of military functions, related industries, and support services.

Jung, Lee, Kim, and Jung (2015) analysed Kure's inheritance of war memory in the context of 'torsion'. They insist Kure did not become a peaceful city, unlike Hiroshima, and regarding that crucial war momentum, Japan's core strength appeared in its post-war US-Japan security. After the Korean War outbreak, the US-Japan alliance was formed, and Kure served as a support base. Also, they argued that 'it is this power that frustrated Kure when seeking to become a postwar peace city' (p. 196) and defeated The Former Military Port Transition Act, aimed at a peaceful city. The next level of Kure's power, which maintained its function as a military port, was the Principle of Inheritance of the Great Japan Empire. In Japanese society, pre-war history is not an object of disconnection, but one of glory, and Kure and Yamato, a symbol of the entire navy, play a big role. As a result, a crack emerged in the norm of the 9th Constitution, the peace constitution, which had defined Japanese post-war society.

Kure has not lost its function as a military port city for 130 years since 1889, except between 1945 and 1954, shortly after the war's end. It remains the main naval port in Japan's Maritime Self-Defence Force, along with Yokosuka. Above all, the Old Naval Building is still used as the Kure District Headquarters. However, while attempting to orient itself as a post-war peace city, Kure had no choice, but to deny its function as a naval port, and the gap created by this negation and the city's actual function created a warped war memory. The space where this torsion is most prominent is in the Yamato Museum.

Additionally, this torsion of war memory is revealed in a strange way by connecting the past and the present. While trying to avoid the memory of former naval divisions, including Kure's ongoing purpose as naval port city, attempts to inherit the glory of the past appeared in the process of registering the four naval cities in the Japan Heritage, a new cultural heritage protection system created in 2015 by the Agency for Cultural Affairs of the Japanese government. As its name suggests, it is a system that selects the 'story' of a region that is intended to be registered as a part of UNESCO's World Heritage. As of 2020, 83 cases have been registered, and unlike conventional cultural heritages, which, for example, seek to preserve a single building, Japan Heritage is usually based on serial registrations of a broad region.

One of Japan Heritage's serial registrations, established on the 25th of April 2016, was 'Chinju-fu (naval base) in Yokosuka, Kure, Sasebo, and Maizuru:

Cities where you can feel the dynamic modernization of Japan'. Interestingly, the aforementioned torsion appears in the title itself. The Japanese and English titles are slightly different: the Japanese version contains 'Naval Base (Chinju-fu)', while the official English title is 'The Four Dynamic Coastal Cities of Yokosuka, Kure, Sasebo, and Maizuru: Centers of Japanese Modernization'. The most important identity of the heritage, the naval narrative, is missing from the English version. The fact that these differences are not simply a matter of translation appears in the context of legacy. The most recent data, a pamphlet issued by the Japan Heritage Utilization Promotion Committee (2018), explained the four cities' heritage values as (originally in Japanese):

> Invited by the sea breeze,
> Time slip to the Meiji era
>
> In Japan, during the Meiji period, four naval cities were established: Yoko-suka, Kure, Sasebo and Maizuru. . . . The naval base (in the Meiji era) was always generously supplied with state-of-the-art industrial technology and equipment. Ports were constructed, where large ships came and went, and steelworks, ships, and tough brick warehouses were improved. . . . In addi-tion, the Navy brought in various cultures, such as new foods, sports, and music. . . . After a period of turbulence and the end of WWII, the four cities were reborn as harbour cities for peace industry, using former military prop-erties. . . . Only these four cities in Japan can share and experience this naval port and naval base history.

As demonstrated earlier, in the four cities' story, as a part of the Japan Herit-age, a direct connection appears between the Meiji period and post-war history. In other words, the naval base was created, and post-war technological development emerged, and these are both instances of Japan's proud legacy. War is represented by the ambiguous term 'a period of turbulence', and in the story, memories of the war from 1910 to 1945 are entirely missing, when the four regions occupied the most important positions in history.

In 2018, the Japanese Heritage also displayed movements that intermittently inherited the memory of the empire through the addition of Kure's constituent cultural assets and the renaming of the naval base. The former basement opera-tion room (former Kure naval headquarters bunker), one of the important warfare facilities, was added, and naval trumpets were renamed to trumpets 'Kimigayo'. Kimigayo (His Imperial Majesty's Reign) represents a Japanese nation that praises kings, symbolising Japanese imperialism, and was the imperial anthem from 1888 to 1945. After the end of the war, it lost its function as an official anthem, but in 1999, it became Japan's official anthem once more through National law.

The contradictory perceptions of the naval base in Japanese Heritage can be perceived as the manifestation of its torsion between acting as a naval port, deviat-ing from its direction towards establishing a peaceful city, and even attempting to inherit the empire.

8.3 Yamato Museum and Japan Maritime Self-Defence Force Kure Museum

The Kure Maritime Museum (i.e. the Yamato Museum) is a symbol of Kure's disconnection and continuity of war memories. As the name implies, it is a museum commemorating the maritime history of Kure, centred on the battleship Yamato.

The ship's construction began at the Kure Naval Arsenal in 1937, the year after Japan's 1936 withdrawal from the London Disarmament Treaty. Commissioned in 1941, it became the world's largest battleship and served as the flagship of the Allied fleet until 1943. However, it did not show any substantial performance as naval warfare moved to the centre of the aircraft carrier, and sunk in the southern part of Nagasaki Prefecture on the 7th of April 1945 during the Japanese Special Attack Units operation.

Apart from its objective achievements, battleship Yamato's images have been reproduced through various media, since the end of the war, with the title of the world's largest battleship, and as a symbol of Japanese technology and the tragic end of the Japanese samurai spirit. The following are some representative works: the novel *Battleship Yamato's End* (1952) by Mitsuru Yoshida, who was a part of the Yamato crew; the movie *Battleship Yamato* (1953), based on that novel; the sci-fi animation *Space Battleship Yamato* (1974), also called *Star Blazers* in English; Henmi Jun's non-fiction *Yamato of Men* (1983), inspired by the records of 100 crew members; the film *Yamato of Men* (2005), based on the above book.

Most of these works used Yamato as a symbol of 'monumental memory' (Han, 2009) of war or as a means of conveying an antiwar message. Kure, the area where the battleship Yamato was built, did not receive much attention in these works. It also did not implement promotions or public relations between the works on Yamato and the region until the early 2000s, because the battleship was treated as a memory of disconnection rather than a continuation with modern Japan.

In Kure, along with the construction of a maritime museum, Yamato, considered a memory of disconnection, appears as the object of succession. According to former Mayor Ogasawara, who was in charge of building the museum in 1991, its first basic concept was created through a discussion on the construction of a marine museum displaying the history of modern shipbuilding technology (Ogasawara, 2007). The issue was that without remembering the navy's technology, the memory of disconnection could not reveal Kure's true history. After the museum's concept was in full swing, its nature gradually changed, as it was revealed that the battleship constructed in Kure during WWII could become its centre.

Among them, the 'Think of Yamato' symposium series held in 1995 for the 50th anniversary of the war was an opportunity for the battleship Yamato to appear in front of the museum. Ogasawara conceived of the idea of a museum designed to make visitors think about war and peace through the history of the navy, and the symposium was put forward as evidence supporting the viability of such a museum. As a result, in 2005, the Kure Maritime History Museum opened under the name Yamato Museum.

Located next to the Kure Ferry Terminal, facing the sea, the Yamato Museum consists of four floors. Some of its attractions include Yamato Square, an exhibit about the history of Kure, a large-scale exhibition hall, a technology exhibition room about crafting ships, and an exhibition room about the future maritime technology.

Visitors are greeted at Yamato Plaza with a 1/10 scale model of the battleship Yamato (Figure 8.1). Although it is said to be a model, it can also be described as the resurrection of the Yamato; it was produced based on actual blueprints from the shipbuilding company in Kure. In Exhibition Room A on the first floor next to Yamato Square, panels and artefacts that show the history of Kure, centred on the navy, are on display. The room shows the process of the creation of the Kure Naval Base and the Naval Shipyard, as well as the process of expanding the economy leading up to the war. An exhibition related to the history of the battleship Yamato follows. It shows the process of building the battleship in Kure, as well as the current state of the Yamato where it sits on the seabed. Photographs and artefacts on

Figure 8.1 1/10 scale model of the battleship Yamato at Yamato Museum.
Source: photo by author

the regeneration of the post-war peace industrial city are presented in the second half of Exhibition Room A.

Located between Exhibition Room A and Yamato Square, Exhibition Room B displays large-scale materials. Most of the artefacts are weapons from World War II, such as human torpedoes and ships (Zerosen) used for suicide attacks (Kamikaze). The museum explains that it wants to convey through these artefacts the misery of war and the importance of peace, but the descriptions of the actual artefacts are devoted to technology, not the horrors of the war.

Exhibition Room C on the third floor is an experiential space where children can learn about shipbuilding techniques. The principle of how the boat floats and various simulators and devices to make a model boat are exhibited. Lastly, Space D on the third floor, labelled 'To the Future', displays the future prospects of ocean and space science and technology and messages from celebrities. What is content is the last space, the Yamato Theatre. Videos about space and ocean exploration are shown, but the central one is the CG video 'Yamato', which explains the construction of the battleship in Kure and how this technology was passed down in Kure after the war.

A road between the Yamato Museum, there is a JMSDF Kure Museum, also known as the Iron Whale Museum. It opened in 2007, two years after the Yamato Museum, but the construction for both was done at the same time. At that time, Mayor Ogasawara asked the Maritime Self-Defence Force to provide exhibits, and in the middle of the lists, there was a real submarine (Ogasawara, 2007). The Maritime Self-Defence Force revised its original plan to donate only submarines for display and created the Maritime Self-Defence Force PR Centre using submarines as part of the exhibition hall. It is the only space in Japan that exhibits actual submarines. JMSDF Kure Museum fills in the missing part of the story told in the Yamato Museum, namely the history of the Japanese naval forces after the war. After the war, Japan's maritime defence was influenced by the Peace Constitution and the U.S.-Japan alliance. Patrol, minesweeping, and coastal submarines occupied a large potion, and the museum's exhibits focus on explaining these parts. The exhibition room on the first floor shows a brief history of the launch of the Maritime Self-Defence Force at the old naval base (devoid of the history during the war) through a panel. That was until 2021, however, and in 2022, the museum made major changes to the content of its ground floor exhibition panels, adding the history of the Imperial Navy from 1868 to 1945. The second section explains the structure and technology of the minesweeper. On the third floor, the submarine's functions and inner workings are shown, and the floor is connected to the retired submarine Akishio on display.

What is interesting about the JMSDF Kure Museum is that an attempt to connect the naval history and the Self-Defence Force during World War II into a continuous history appears through special exhibitions, titled 'Ship X Ship' held in 2018 and 'Ship X Ship II' held in 2019. Many ships attached to the post-war Marine Self-Defence Force have inherited the ship names used in the Naval Period, and the two exhibitions show that the pictures of the ships in the Navy era and the photos of active ships are matched and connected. In other words, it can be said that it is a work that inherits wartime history as a continuity. However, The JMSDF's attempt to inherit the memory of the war with this continuity is dramatised in the aforementioned change in the content of the exhibition panels in 2022. The succession of war memories and the memory of the Imperial Navy, which had been presented in an oblique form, will now be represented in a more formal way.

Through the Yamato Museum and the JMSDF Kure Museum, Kure's maritime history became a more contradictory form of both disconnection and continuity. Uesugi (2012) discussed this concept as follows:

> When Kure reflects on the previous history of technology used for peaceful means, it comes to a realization that the navy must be excluded, and that war is evidently a threat to peace. On the other hand, based on the concept of the 'respect for peace,' which came before and after the war, if the technology, cultivated before and after the war, developed and reached present day science and technology, "respect for peace's" meaning itself would be in crisis.
>
> (p. 130)

On the other hand, Jung et. al. viewed the relationship between Yamato and Kure not as a local narrative but through a larger perspective, because Kure is a symbolic place for Japan, where a rightward drift has been maintained since the 2000s. Specifically, this torsion refers to the contradictions between US-Japan security, the Great Japanese Empire's inheritance principle, and Article 9 of the Peace Constitution.

8.4   De-contextualisation and re-contextualisation of the battleship Yamato

One of the new trends that emerged around inheritance of war memories related to battleship Yamato in recent years is de-/re-contextualisation. It has been well represented through popular culture since the 2000s. The Japanese sociologist Azuma (2001) calls this phenomenon animalisation and database consumption. The popular culture consumption of the 1970s, when *Space Battleship Yamato* and *Mobile Suit Gundam* were created, connected the individual society to the great world view. On the other hand, since the 2000s, popular culture enthusiasts, especially the otaku, tend to only consume characters and representations, without linking the entertainment to society.

Against this backdrop, de-contextualised popular cultural works and their non-contextual consumption, which are based on World War II, have been on the rise since 2010. For example, in many cases, these works personify weapons related to World War II as beautiful females, as they are tailored to young males, the main consumers. Many fans relate war weapons and connected places in the works to their favourite female characters in the game. Indeed, we are interested in and search for individual historical information, but even character consumption is often separated from war memory. In other words, some fans are truly interested in finding individual war-related historical information, but this content is often separated from actual war memories.

A representative example of non-contextual consumption of de-contextualised popular culture works related to battle ship Yamato is the Japanese PC game *Kantai Collection* (aka *KanColle*), which was released in 2013. It is an anthropomorphic game that embodies Japanese warships as female characters (Kanmusu) during World War II and aims to collect various battleship characters and develop their ranks to admirals. Initially, only Japanese World War II warships appeared, but the game gradually added not only Axis but also Allied warship characters. Military and beautiful girl characters are popular in Japan, and the game has gained great popularity since its initial launch to the present day. On the other hand, since it is a game about the Japanese navy during World War II, a controversy emerged over right-wing propensity, and for this reason, it is not officially available in Korea, China, and Taiwan. However, a large number of individuals from the younger generations still enjoy playing the game in those three countries and have acquired it through various detours.

KanColle's distinctive feature is that the characters intervene, and replace the history and memory of the war with casual consumption objects, which further extends to the non-contextual consumption of Yamato Museum, Kure, and other war-related places. For example, fans who visited the Kure naval cemetery put *KanColle* character cards on almost every memorial monument of battleships, including Yamato. This behaviour may be viewed as inappropriate in Japanese society. As such, social media and other fans criticised it as being unreasonable.

On the other hand, numerous sources present concerns regarding fans' need to grasp the historical truths contained in the non-contextual consumption of de-contextualised war-related pop culture. Accordingly, a South Korean columnist commented on *KanColle*:

> The controversy of the *Kantai Collection*, coupled with the de-political debate in games, has made the Internet buzz. It is difficult to conclude whether the game intentionally beautified the history of the Japanese invasion, but fans must know the historical truth that permeates the content. How much pain have the weapons used in Japanese aggression brought us?
>
> (Game About, 2015)

However, more important than the game's context is that de-populated war-related pop cultures are re-contextualised, especially with political intent. De-contextualisation means that there is room for any form of context to fill the void, which allows for various re-interpretations and meanings.

A representative example occurred during the 130th anniversary of the Kure Naval Base (130th KURE 2019), a collaborative event between Kure and Kankore Production held in October 2019, with two-day night dances, special live performances by voice actresses, a stamp rally, and joint events with the Yamato Museum, the Irifuneyama Memorial, and local restaurants. The event filled the gap between Meiji and the post-war era. In principle, Kure was reborn as a peaceful industrial port city, which the Japanese Heritage emphasised. On the other hand, in real history, Kure is still a major naval port in Japan. In this situation, Kure neither denies the disconnect nor actively claims the continuation of war memories. Among these, *KanColle*, a popular culture computer game, was able to play a role in connecting the two smoothly.

8.5 Conclusion

In this chapter, focusing on the Yamato Museum, I looked at how the memories of war are inherited and reproduced in the midst of disconnection and continuity. The Yamato Museum represents the inheritance of war memories at the local level while also acting as a miniature version of Japan's war memory succession at the national/global level. It shows how the attempts to disconnect the war memories, which were made immediately after the end of the war, created a contradiction in the regions that inherited it. On the other hand, Article 9 of the Constitution, which is a large framework that penetrates the entire post-war Japanese society, did not work equally throughout all of society; it created cracks in which the normative power was limited. Paradoxically, one of the reasons for this crack was the U.S.-Japan security treaty created to deter war in Japan. The Yamato Museum shows the contradictions and gaps in the inheritance of war memories in post-war Japan. Among them, attempts to connect war memories with current politics and national buildings have emerged. As can be seen from the Yamato Museum and the JSMDF Museum, small attempts are being made to neutralise the normative power of Article 9 of the Constitution.

On the other hand, the Yamato Museum has the function of representing the memories of war that are de-contextualised and re-contextualised with the passage of time. The Yamato Museum functions as the basis for the re-contextualisation of war memories that have lost context, while memories of World War II are reduced to casual consumption objects through popular culture, which again acts as a means of re-contextualisation at the political level. The post-contextual pop culture and the local history, as well as inheritance of war memories, create an ambiguous contact point, moving back and forth between fans and the Yamato Museum.

Like the hull of the battleship Yamato, which sank into the deep sea with only a salvage plan in place, Kure's war memories are still struggling between disconnection and continuity.

References

Allen, M., and Sakamoto, R. (2013). War and peace: War memories and museums in Japan. *History Compass*, 11–12, 1047–1058.

Azuma, H. (2001). *Otaku: Japan's database animals*. Minneapolis, MN: University of Minnesota Press.

Dower, J. W. (2012). *Ways of forgetting, ways of remembering: Japan in the modern world*. New York: The New Press.

Frost, M. R., Vickers, E., and Schuhmacher, D. (2019). Locating Asia's war memory boom: A new temporal and geopolitical perspective. In M. R. Frost, D. Schumacher, and E. Vickers (Eds.), *Remembering Asia's World War two* (pp. 1–24). London: Routledge.

Game About. (2015). *Abe and the Japanese game that follows him 'right-handed'*. https://post.naver.com/viewer/postView.nhn?volumeNo=2546219&memberNo=11710666.

Han, J. (2009). "Jeonhu ilbon-ui ginyeombijeog gieog: Manhwayeonghwa<ujujeonham yamato>wa 1970nyeondae jeonhusedae" (Monumental Memories in Postwar Japan: The Animation <Space Battleship Yamato> and the Postwar Generation of the 1970s), *Sahoe-wayeogsa (Society and History)*, 83, 83–113.

Japan Heritage Utilisation Promotion Committee. (2018). *Around the Japanese heritage of 'Chinjufu'*. Pamphlet.

Jung, G., Lee, H., Kim, M., and Jung, Y. (2015). *Powidoen pyeonghwa, guljeoldoen jeonjaeng gieog: Hilosima man-ui gunhangdosi gule yeongu* (Surrounded peace, refractory war memories: A study on Kure City, Hiroshima Bay). Seoul: Jeiaenssi.

Lee, J. (2018). Yasukuni and Hiroshima in clash? War and peace museums in contemporary Japan, *Pacific Focus*, 33(1), 5–33.

Ogasawara, S. (2007). *Senkan 'Yamato' no hakubutsukan: Yamato myūjiamu no tanjō no zen kiroku* (Battleship Yamato Museum: Complete record of the birth of the Yamato Museum). Tokyo: Fuyōshobō shuppan.

Uesgi, K. (2014). Gunkō toshi <Kure> kara heiwa sangyō kōwan toshi <Kure> e (From naval port city <Kure> to peace industry port city <Kure>). In Y. Sakane (Ed.), *Chiiki no naka no guntai 5: Nishi no guntai to gunkō toshi Chūgoku Shikoku* (Regional troops 5: Western area and the military port city of Chugoku and Shikoku) (pp. 104–130). Tokyo: Yoshikawakōbunkan.

Uesugi, K. (2012). Renzoku to danzetsu no toshi-zō: Mōhitotsu no "heiwa" toshi Kure (Urban image of continuity and disconnection: Kure, another "peace" city). In Y. Fukuma, M. Yamaguchi, and K. Yoshimura (Eds.), *Fukusū no Hiroshima: Kioku no sengo-shi to media no rikigaku* (Multiple Hiroshima: Postwar history of memory and media dynamics) (pp. 103–138). Tokyo: Seikyusha.

Winter, J. (2001). The memory boom in contemporary historical studies. *Raritan*, 21(1), 52–66.

Yamane, K. (2017). Japanese peace museums: Education and reconciliation. In E. Hunter (Ed.), *Peace studies in the Chinese century: International perspectives* (pp. 85–113). New York: Routledge.

Yoneyama, L. (1999). *Hiroshima traces: Time, space, and the dialectics of memory*. Berkeley, CA: University of California Press.

9 'Kamikaze' heritage tourism in Japan

A pathway to peace and understanding?
(*Journal of Heritage Tourism,* Vol. 15, 2020, Issue 6, 709–726)

Richard Sharpley

9.1 Introduction

Over the last two decades, Japan has witnessed remarkable growth in inbound tourism. In 2000, 4.7 million international arrivals were recorded; by 2010, this figure had almost doubled to 8.6 million but, most notably, an average annual growth rate of around 28 per cent was achieved between 2012 and 2017 (JNTO, 2019a). By 2018, international arrivals totalled almost 31.2 million (JNTO, 2019b). This rapid and sustained increase in international tourism reflects a deliberate policy on the part of the Japanese government, the principal objective of which is economic growth and regional revitalization (MLIT, 2016: 3), not least to address the challenge of an ageing and declining rural population (Crowe-Delaney, 2019). At the same time, however, it is also evidence of what Funck and Cooper (2015: 46) refer to as Japan's 'long-cherished role of fostering long-standing friendship and trust among nations', more recently formalized in its 2013 National Security Strategy (Oros, 2015). Indeed, an explicit objective of promoting tourism is the 'Enhancement of mutual understanding. . . . To raise the current position of Japan our forerunners achieved in the condition of international peace [*sic*] and to accomplish our responsibilities at present and in the future' (MLIT, 2012: 4). A more recent policy accords greater significance to economic priorities yet still views tourism as a means to 'foster dynamic multicultural exchange' (MLIT, 2016: 3), including, more pragmatically, improving visitor access to and the presentation of heritage sites (MLIT, 2016: 9).

Yet, the extent to which such 'mutual understanding' can be achieved is debatable. On the one hand, 'Japan today projects an external image in which harmonious coherence provides a basis for technical efficiency and cultural excellence' (Pye, 2003: 45), whilst the contemporary hosting of major international events, such as the successful Rugby World Cup in 2019 and the Summer Olympics in Tokyo (postponed to 2021), points to positive symbiotic engagement with the international community. On the other hand, peaceful international (and indeed, domestic) relations are considered by some to remain challenged, not least by the manner in which Japan confronts its twentieth-century military past. This is not to say that the country does not seek to promote peace through its wartime heritage.

DOI: 10.4324/9780367823795-13

Notably, since it was designated by the government as 'Japan's International City of Peace' in 1949 (Yoshida, Bui & Lee, 2016), Hiroshima has not only taken on the leadership of a global anti-nuclear weapon movement but also developed a successful 'peace tourism' industry based upon its tragic past (Schäfer, 2016) and is emblematic of a contemporary Japan adopting a proactive role in international peace and prosperity (Oros, 2015). However, as some argue, Hiroshima primarily encourages a victim consciousness (for example, Siegenthaler, 2002), whilst, as the author of this paper has personally observed, the newly refurbished Hiroshima Peace Memorial Museum reveals powerfully the destructive effects of the A-bomb but little attempt is made to explain *why* it was dropped on the city. Hence, commentators such as Takenaka (2015: 16) argue that despite increasing manifestations of official remorse for the country's war-time activities, Japanese society today feels 'no obligation to engage in the post-war responsibility discourse', whilst others suggest that, as at Hiroshima, a dominant domestic narrative of victimhood shapes contemporary attitudes toward the county's military past and its approach to international relations (Orr, 2001; Yasuaki, 2002). Either way, however, it is evident from the literature that the representation of Japan's wartime heritage is both varied and controversial (Allen & Sakamoto, 2013); as Nelson (2003: 445) summarizes, 'since the end of World War II, diverse interpretations over how to represent, acknowledge and atone for Japan's aggressive exploits throughout Asia and the Pacific have occasioned as much controversy and conflict as they have closure'.

Much of this controversy relates to Yasukini Jinja (Shinto shrine) in Tokyo, where more than 2.5 million Japanese war dead, including a number of A-Class war criminals, are not only commemorated but apotheosized (Breen, 2004). Consequently, the shrine remains a source of both national and international political conflict. Yasukuni is referred to again later in this paper but a more specific and equally controversial issue is the commemoration of so-called kamikaze pilots, more formally referred to as *tokkō*, or Special Attack Force (Sheftall, 2008: 155), who undertook suicide missions during the last year of the Pacific War. (It should be noted that, as discussed later, suicide missions were also undertaken by divers, suicide boats, and 'human torpedoes', or *kaiten*. However, the focus of this paper is on kamikaze aircraft pilots.) Between October 1944, when the first officially sanctioned kamikaze operation took place against the American fleet approaching the Philippines (Axell & Kase, 2002: 40), and the formal end of the war on 2 September 1945, some 3,000 Japanese airmen died in kamikaze attacks (though some put the figure as high as 5,800 – see Sheftall, 2008). They are now commemorated at a number of museums and shrines around Japan and, unsurprisingly, a number of studies explore the kamikaze phenomenon and its contemporary representation (Allen & Sakamoto, 2013; Axell & Kase, 2002; Inuzuka, 2016; Nelson, 2003; Sakamoto, 2015; Sheftall, 2008; Yoshida, 2004). Typically, however, these focus on the kamikaze within competing discourses on war heritage within Japan; in contrast, the potential role of kamikaze heritage sites in both enhancing what is arguably limited knowledge of the phenomenon and encouraging wider international understanding and reconciliation through tourism has not been considered. This is a notable omission, not least given that, in addition to the broader objective of promoting mutual understanding, the most recent tourism policy in Japan targets Western nations, in particular the United States, as key international tourist markets (MLIT, 2016).

This chapter, therefore, seeks to address this gap in the literature. Specifically, it explores the manner in which two heritage sites in Japan, namely the Chiran Peace Museum in Kagoshima Prefecture (the country's principal heritage site dedicated to kamikaze pilots) and the Yūshūkan War Museum located within the grounds of Yasukuni Jinja in Tokyo, commemorate and interpret the exploits of the kamikaze pilots for both domestic and, in particular, international tourists. To do so, it is necessary to consider the historical context in which the kamikaze attacks were conceived, authorized, and undertaken, but the first task is to review how tourism may, in principle, play a role in achieving peace, understanding, and reconciliation, particularly in the context of post-conflict heritage sites, as a framework for the subsequent discussion.

9.2 Tourism, peace and mutual understanding

It is observed that, pragmatically, 'tourism is far more dependent on peace than peace is on tourism' (Hall, Timothy & Duval, 2004: 3); indeed, research has concluded that although international tourism may in some circumstances foster cross-cultural understanding, typically 'tourism is the beneficiary of peace rather than grounds for peace' Pratt & Liu, 2016: 82). Nevertheless, it has long been suggested that tourism can be a pathway to international peace and mutual understanding. Following the First World War, travel was promoted to encourage peaceful relations between former adversarial nations whilst in 1967, the United Nation's International Tourism Year was themed 'Tourism: Passport to Peace'. Subsequently, the World Tourism Organization (WTO, 1980) identified international tourism as a vital force for peace, an objective carried forward by D'Amore (1988), who, in 1986, founded the International Institute for Peace through Tourism. Since then, increasing attention has been paid to the relationship between tourism and peace (Blanchard & Higgins-Desboilles, 2013; Moufakkir & Kelly, 2010), the belief being that social and cultural connections through tourism 'spur dialogue and exchange, break down cultural barriers and promote the values of tolerance, mutual understanding and respect' (Rifai, 2013: 11).

Inevitably, such an overarching ambition for a phenomenon as extensive and diverse as contemporary tourism may be considered unrealistic (Litvin, 1998). Certain, particularly mass, forms of tourism are unlikely to offer opportunities to encourage peace and understanding (Harrison & Sharpley, 2017); moreover, the increasing evidence of so-called overtourism (Milano, Cheer & Novelli, 2019; Pechlaner, Innerhofer & Erschbamer, 2020) suggests that, increasingly, conflict and resentment between tourists and destination communities may actually be the outcome of tourism development. At the same time, as Farmaki (2017) observes, research into the tourism and peace nexus tends to be based upon the contact hypothesis – that is, it is assumed that contact between tourists and relevant members of destination societies provides the basis for developing mutual understanding and promoting peaceful relations. This, she argues, is simplistic; in order to assess the potential for reconciliation and peace through tourism, it is necessary to take into account a variety of factors including the initial cause of the conflict and contemporary contextual influences that may inhibit its resolution, as well as the nature, role and governance of tourism in the destination. In the context of this paper, such contextual influences might include competing perspectives on how to commemorate Japan's war dead (Jeans, 2005; Yoshida, 2004), a long-standing sense of victimhood dating back to the politics of the immediate postwar years (Tsutsui, 2009) and contemporary shifts in what Oros (205) refers to as the country's security identity.

Nevertheless, it is acknowledged that in some contexts, particularly at tourist sites related to difficult pasts – or sites of difficult heritage (Logan & Reeves, 2008) – the opportunity exists, through appropriate commemoration and interpretation, to encourage reconciliation and understanding. More specifically, heritage sites of, or related to, conflict between peoples and nations, such as battlefield sites, museums dedicated to violent pasts, sites of genocide or memorials to those who have lost their lives in acts of terror or violence, represent a legitimized space where, in principle at least, visitors with direct, indirect or even no connection with the event, its victims and indeed its perpetrators may congress to both remember those who suffered in a violent past and also to reconcile differences in order to establish a more peaceful present and future (Gurler & Ozer, 2013), to ensure that 'never again' can such events occur (Lollis, 2014; Williams, 2007). In short, post-conflict heritage sites may foster accountability, justice, debate and reconciliation, or the transformation of 'relations of hostility and resentment to friendly and harmonious ones' (Bar-Siman-Tov, 2004: 4; Friedrich, Stone & Rukesha, 2018), as a necessary foundation for peace-building (Buckley-Zistel & Schaefer, 2014).

The extent to which this might occur inevitably depends on a number of factors, including those proposed by Farmaki (2017) referred to above. On the one hand, the opportunity exists at such sites to inform, educate and present 'the truth', in so doing encouraging all stakeholders to reflect, communicate and overcome past differences in a spirit of tolerance and forgiveness. However, a number of requirements should necessarily be fulfilled, not least that the site be accessible and welcoming to all who wish to visit, and also be managed and presented in a manner that fosters reconciliation, including the appropriate or accurate representation of all stakeholders' heritages or stories (Kelly & Nkabahona, 2010). Moreover, there should ideally exist both acknowledgement of the need for and a wider culture of reconciliation; in other words, all stakeholders should from the outset be committed to achieving such peace and understanding (Friedrich, Stone & Rukesha, 2018).

On the other hand, the very nature of difficult heritage challenges its effective or appropriate representation and commemoration. First and foremost, it is highly susceptible to political influence (Sharpley, 2009). In other words, the development and interpretation of difficult heritage may be undertaken to convey particular political messages or an authorized heritage discourse (Smith, 2006). It has long been recognized that, owing to its significance and visibility as a social and economic phenomenon, tourism generally may be exploited for political or ideological purposes (Richter, 1983), most usually to affirm or strengthen cultural identity. As Cano and Mysyk (2004: 880) observe, 'the state may assume the role of marketer of cultural meanings, in which it attempts to make a statement about national identity by promoting [through tourism] selected aspects of a country's cultural patrimony' (see also Wight, 2016). Such a statement may be intended primarily for a domestic audience (Palmer, 1999); equally, it may also be intended for international tourist consumption, such as at the Kigali Genocide Memorial in Rwanda, where the interpretation of the 1994 Genocide and its aftermath arguably seeks to legitimize the contemporary government of the country to international visitors (Sharpley & Freidrich, 2016).

The outcome of political intervention in difficult heritage is but one potential source of dissonance or dissonant heritage. As Tunbridge and Ashworth (1996: 21) observe, 'all heritage is someone's heritage and therefore logically not someone else's, and the original meaning of an inheritance implies the existence of disinheritance'. This disinheritance occurs when there exists a 'lack of congruence at a particular time or place between people and the heritage with which they identify' (Ashworth & Hartmann, 2005: 253), or, alternatively stated, dissonant heritage is manifested when the past is represented or interpreted in such a way that, for particular people or 'subaltern' (Smith, 2006: 35) groups, their past is distorted or displaced. All heritage is inevitably susceptible to dissonance, not only because of the existence of multiple interest groups but also because, over time, people's relationship with and understanding of past events may change. However, that dissonance may be enhanced when the representation of past events '(re)interpret[s] past events to meet contemporary political agendas, to erase or deny a particular past [or] to celebrate victory' (Sharpley, 2009, 150–1). Moreover, Ashworth and Hartmann (2005: 254) argue that not only does 'human tragedy imbue [dissonance] with a capacity to amplify the effects and thus render more serious what otherwise would be dismissable as marginal or trivial' but also that dissonance is inevitable in difficult heritage construction and interpretation.

In the context of this paper, the existence of dissonance at difficult heritage sites in general and at post-conflict sites in particular is, by definition, likely to hinder the promotion of understanding and reconciliation. Specifically, unless the heritage of all those involved or with an interest in a former conflict is recognized and fully addressed at post-conflict sites, the necessary foundations for peace-building referred to by Buckley-Zistel and Schaefer (2014) will not be in place. And as argued below, this is indeed the case at the two kamikaze heritage sites in Japan. Not only do they present an authorized revisionist narrative of the heroic sacrifice of the kamikaze pilots' exploits but little or no attempt is made to locate that narrative within the wider political and cultural context of the time, a context that might foster a more nuanced understanding of the kamikaze phenomenon amongst both domestic and international visitors. Therefore, before discussing the interpretation and commemoration of the kamikaze at Chiran Peace Museum and Yūshūkan War Museum, the following section offers an overview of the historical and cultural circumstances surrounding the kamikaze phenomenon.

9.3 The kamikaze: the historical context

The word 'kamikaze' is arguably imbued with a certain historical romanticism. Translating as 'divine wind', it was the name given to a powerful typhoon that, in 1281, destroyed the invading fleet of the Mongolian Emperor Kublai Khan as it approached Japan, leading the Japanese to believe that, if once again facing attack, their country would be similarly saved by a divine wind (Chiran, 2017: 11). However, contemporary perceptions of the kamikaze phenomenon vary both within and beyond Japan. It is suggested, for example, that to younger Japanese it 'is a curiosity from the past; to the older it is a reminder of a cruel epoch' (Axell & Kase, 2002: 3); more specifically, some Japanese consider the kamikaze pilots to have been 'irrational, heroic and stupid' (BBC, 2017). To non-Japanese, in contrast, it is probable that many maintain a perception of kamikaze pilots as 'men willing, almost gladly, to die in the name of their country and for the sake of their emperor . . . a soldier in a cockpit, ready to do his duty, piloting his plane into the deck of an American ship . . . and in death, becom[ing] something noble' (Konstantopoulos, 2007: 6). Either way, however, the narrative surrounding the phenomenon remains highly contested within Japan (Jeans, 2005; Sheftall, 2008) and, in all likelihood, misunderstood beyond the country's borders. Nevertheless, the historical events leading up to the official commencement of kamikaze attacks in 1944 are generally acknowledged, as are the broader cultural influences on the apparent willingness of the young pilots to undertake the missions.

9.3.1 The kamikaze strategy

The Pacific War commenced on 8 December 1941 when, following the attack on Pearl Harbour, Britain and the USA declared war on Japan. The events leading up to the attack are complex (Costello, 1982). However, it is generally considered to have been the inevitable outcome of Japan's foreign policy, specifically its 1931 invasion and occupation of Manchuria (north-east China), leading some to consider the Pacific War as part of the wider Asia Pacific War of 1931–45 (Ienaga, 2008; Takenaka, 2015). Japan invaded Manchuria primarily to exploit the region's natural resources but also to fulfil the broader intention of establishing a 'New Order in East Asia' (Axell & Kase, 2002: 21). This resulted in growing opposition amongst Western nations to Japan's international activities, culminating in the USA imposing, amongst other things, an embargo on the export of iron and aviation fuel to Japan. As a country with few natural resources of its own, this was seen as a threat to not only its international ambitions but its very survival, although the commencement of hostilities against the USA in 1941 and subsequent occupation of a number of territories in the region was also ironically seen by Japan as an opportunity to liberate Asia from the control of the world's then Western colonial powers.

Initially, the country enjoyed a number of military successes; however, by mid-1942, the tide began to turn. In particular, the Battle of Midway in June of that year, in which much of Japan's naval air power was lost, was a turning point in the Pacific War and her armed forces began to suffer a number of setbacks. By mid-1944 it had become recognized by Japan's military leaders that the continuing advances of numerically superior American forces could no longer be countered by conventional tactics and, hence, with the Japanese-occupied Philippines facing imminent attack, and in the face of overwhelming odds, the first Special Attack (*tokkō*) missions were launched against American ships in the October of that year. Subsequently, the kamikaze campaign reached its peak near Okinawa during the following summer, finally ending with the cessation of hostilities in August 1945.

It should be noted that the decision to authorize the Special Attack missions, most famously the kamikaze airstrikes but also utilizing manned torpedoes, or *kaiten*, suicide boats (*shinyo*) and suicide divers, was not universally accepted by Japan's military leaders. Particularly, some questioned the rationale of sending men to certain death but with less certainty with regard to successful outcomes (Axell & Kase, 2002); for example, only an estimated 10 per cent of kamikaze pilots actually crashed into their targets (BBC, 2017), whilst *kaiten* missions are believed to have sunk only three American ships at the cost of 106 *kaiten* pilots, many of whom died on training missions or because of equipment failure (NHHC, 2019). Equally, there is evidence to suggest that, for a variety of reasons, many of those who flew on kamikaze missions were not in fact willing volunteers (Sheftall, 2005). Some deliberately ditched their aircraft in the ocean rather than crashing into ships, whilst many of those who survived (the war having ended prior to their missions taking place) subsequently revealed their unwillingness to participate (Axell & Kase, 2002; BBC, 2017). However, not only did the process of 'volunteering' make it difficult to refuse but also the pilots 'knew there was no alternative. To refuse to fly was to show a lack of duty to their country and their parents' (Chiran, 2017: 19), not least because of the tradition of honourable death, based upon the notion of *bushido*, that was deeply entrenched in the culture of the Japanese military.

9.3.2 Bushido

The concept of *bushido*, which translates literally as 'military knight ways' (Nitobé, 1908), can be traced back to the period of the Meiji Restoration from 1868. Its roots, however, lie in Tsunetomo Yamamoto's *Hagakure* (see Yamamoto, 2002). Written in the seventeenth century, this described the morals and ethics of the Samurai in general and, in particular, the belief that it was better to achieve 'one's aim in death . . . than a continued failure to do so in life' (Konstantopoulos, 2007: 11). The *Hagakure* was subsequently adopted by the early Meiji government as a means of unifying Japan – that is, as a means of transferring the loyalty held by the former Samurai warrior class to their feudal lords to a sense of loyalty to their emperor amongst the Japanese people more widely in the newly unified Japan.

In essence, then, *bushido* represents a moral code, a set of principles that the ancient Samurai were required to observe in all aspects of their lives, including justice, courage, benevolence, politeness, truthfulness, honour and loyalty (Nitobé, 1908). In his widely cited essay, Nitobé (1908) explains that *bushido* is underpinned by the twin influences of Buddhism and Shintoism. On the one hand, Buddhism encourages a sense of trust in fate, acceptance of the inevitable and stoic composure in the face of adversity. On the other hand, Shintoism promotes loyalty above all to the emperor, filial piety (respect for parents and elders) and, in particular, reverence for ancestral memory inasmuch as it is believed that the souls of the dead remain behind to watch over the living. Specifically, in Shintoism, those who die tragically or heroically can be worshipped as gods and therein can be found the second way in which young Japanese pilots were encouraged to die for their country, namely, through the development of the myth surrounding Yasukini.

9.3.3 *The myth of Yasukuni: noble enshrinement*

As noted earlier in this paper, Yasukuni Jinja in Tokyo is the focus of considerable controversy that relates broadly to the relationship between 'the post-war Japanese state and the war dead' (Breen, 2004: 76). On the one hand, there are those who claim that Yasukuni, where the nation's war dead are apotheosized, is a purely religious institution which should not be formally visited by the head of state. To do so is in contradiction to the separation of state and religion as enshrined in Japan's post-war constitution. This position is challenged, on the other hand, by those who argue that the state should rightfully honour its war dead; to not do so, it is argued, is to succumb to external political pressure. More specifically, however, it was the enshrinement of 14 Class-A war criminals at Yusukuni in 1978 and subsequent visits by a number of Japanese prime ministers that have fuelled this controversy (Inuzuka & Fuchs, 2014).

A full consideration of the debate is beyond the scope of this chapter (see, for example, Okuyama, 2009; Pye, 2003; Ryu, 2007; Shibuichi, 2005). Importantly, however, the contemporary controversy surrounding Yasukuni is the outcome of a process, commencing with the Shrine's establishment in 1869, through which not only dying for the Emperor in battle came to be seen in Japan as 'an act worthy of aspiration and a source of pride' (Takenaka, 2015: 2) but also the war dead became the 'protector god for Japan' (ibid., 27). This process is referred to by Takenaka (2015: 26) as the development of the Myth of the War Experience in which death in battle came to be seen as both sacrifice and resurrection; moreover, fundamental to this process was the adaptation of traditional death rituals in Japan.

At risk of simplification (see Takenaka, 2015 for a detailed discussion), in such rituals the founding ancestor of a family's lineage (or *ie*) was considered the family god who protected the living and, thus, was annually commemorated. Other family members who subsequently died would be commemorated for a prescribed period until their spirit was considered to have merged with that of the family god. Over time, however, this ritual was appropriated so that the emperor came to be seen as the founding father of the family, which, by extension, was the nation. As a consequence, the responsibility for commemorating war dead transferred from the family to the nation (at Yasukuni) and their spirits collectively became the god of Yasukuni, protector of the nation. Significantly, by the time of the Asia Pacific War, the spirits of the war dead at Yasukuni were also referred to collectively as *eirei*, which, according to Takenaka, 2015, 90–93), is an invented term meaning noble spirits. Through the enshrinement process, war dead were cleansed of any wrong-doing during their lifetime, hence the justification for the inclusion of Class-A war criminals at Yasukuni and, arguably, the continuing limited sense of responsibility in contemporary Japanese society for their country's war-time aggression.

In short, along with following the code of bushido, kamikaze pilots were, in a sense, offered the opportunity to not only sacrifice themselves for the nation and bring honour to their families but also become noble spirits collectively protecting the nation. And it is this cultural context of *bushido* and promised deification that goes some way to explaining the dissonant nature of their commemoration (and consequential implications for encouraging mutual understanding amongst all visitors) that is now discussed.

9.4 Kamikaze heritage interpretation in Japan

In order to explore the manner in which it is presented and interpreted for touristic consumption, visits were undertaken by the author to two of the country's principal kamikaze heritage sites: Chiran Peace Museum and Yūshūkan War Museum. At each site, field notes and, where appropriate, photographs were taken. In addition, secondary data sources, such as visitor comment books, English language guide books and related extant research, were drawn upon, and relevant online sources were also accessed. Collectively, these facilitated a critical interpretative analysis of the messages conveyed by each museum.

9.4.1 Chiran Peace Museum

Chiran Peace Museum is located near the site of a former air base at the southern end of the island of Kyushu from where, in the final months of the Pacific War in 1945, many kamikaze missions were launched against the American fleet at Okinawa. In fact, 'more kamikaze pilots took off from Chiran than anywhere else' (Chiran, 2017: 5). Little evidence of the airfield remains; however, the Chiran Resource Centre was initially constructed by the town in 1975 to preserve and display memorabilia, such as letters, photographs and others materials, from the kamikaze operations. In the mid-1980s, the centre was redeveloped with state funding and re-opened in 1987 as the Chiran Peace Museum. Commemorating 1,036 kamikaze pilots (their average age was 21.6 years), almost half of whom flew out of Chiran, the museum now attracts more than 500,000 domestic visitors annually, as well as around 10,000 international visitors.

The dominance of the domestic market is not, perhaps, surprising given the museum's location, it being relatively distant from the typical international tourist itinerary in Japan. However, it is clear that from both a practical perspective and from the message it conveys, the museum is also designed primarily for the domestic audience. Regarding the former, an audio-tour in English is available but, with the exception of a sign in the main entrance hall and a small number of (arguably, carefully selected) translated letters from pilots to their families, all information and interpretation is in Japanese. Consequently, and as lamented by many (see

Tripadvisor, 2020), the experience of international visitors is limited to and pre-scribed by the narration on the audio-guide. Notably, this commences by informing visitors that the museum refers to the pilots as *tokkō* rather than kamikaze because they attacked only military and not civilian targets. In so doing, an attempt is argu-ably being made not only to legitimize the kamikaze strategy but also, perhaps, to distinguish it from the more recent Islamic suicide bombings with which commen-tators on the kamikaze phenomenon sometimes draw comparisons (Axell, 2002).

With regard to the latter, on approaching the museum building, visitors first encounter a full-sized replica of a kamikaze plane. This was used in the 2007 Japanese war movie *For Those We Love*, written by Shintaro Ishihara, the then right-wing governor of Tokyo (Danielsen, 2007), which celebrated the heroism of the kamikaze pilots. As such, it reflected the nationalistic and revisionist narra-tive of other recent Japanese war movies although, according to Danielsen (2007), unlike the popularity of other movies, it 'proked disquiet' amongst audiences, not least for parallels drawn with contemporary ideology-driven suicide bombers. Nearby are the statues of both a kamikaze pilot (Figures 9.1 and 9.2) and of Tome Torihama, who owned a restaurant in the town where pilots would go, sometimes with their families, prior to departing on their final mission. She became known as *tokko no haha*, or the 'mother of special attack pilots' (Inuzuka, 2016: 151), and also featured in *For Those We Love*. Her restaurant is now a small museum. These installations immediately establish the focus of the museum on the pilots them-selves and, on entering the main building, this is confirmed when visitors, are first confronted with a large mural (the 'Chiran Requiem') of a pilot being carried from his burning plane by angels.

Figure 9.1 Replica kamikaze plane, Chiran Peace Museum.
Source: photo by Richard Sharpley

Figure 9.2 Statue of a kamikaze pilot, Chiran Peace Museum.
Source: photo by Richard Sharpley

This serves to both represent the deification of the kamikaze pilots at Yasukuni and their honourable death but also, more significantly, sets the overall tone of the museum's perspective on the kamikaze phenomenon. In front of the mural, a sign (referred to above, in English) states that 'the Peace Hall was built here in commemoration of the pilots who died heroically in the skies and to impart the historical realities behind their lives, as well as pray for enduring peace'. Next to the mural, a continual video shows original footage of the kamikaze missions, emphasizing the challenges the pilots faced in reaching their targets. Many planes are seen crashing into the sea, but successful attacks are also shown. Also, in an adjacent room, physical items are on display, including the wreckage of a plane recovered from the seabed off the coast of Kagoshima. These, along with other artefacts on display elsewhere in the museum, such as clothing and other equipment used by the pilots, are tangible and uncontroversial exhibits. They include a replica of an

A-frame barracks hut, located between the museum and a modern shrine dedicated to them, in which pilots would have been billetted prior to their missions.

Most of the museum, however, is given over to photographs of each of the 1,036 pilots commemorated. Accompanying each photograph is a description of the individual's background, age, education, family and military experience. In addition, original copies of their final letters to family and loved ones can be seen in display cases beneath the photographs although, as noted earlier, with the exception of a small number of such letters displayed together in a separate section of the museum, these are not accompanied by translations into English. However, a selection of translated letters is provided in the museum's official booklet on the kamikaze phenomenon, *The Mind of the Kamikaze* (Chiran, 2017), available for purchase. Reproduced under themes such as 'The Mothers', 'Love for Children' and 'Friendships', these letters collectively convey a message of duty, selfless willing sacrifice, of respect and love for parents and family and happiness for dying for the emperor and country. For example, one letter says simply: 'Dear Parents and everyone in my family, At last the long awaited chance has come. It will be my honor to descend into the ocean with my enemy' (Chiran, 2017: 26). Another says, 'Dear Mother and Brother, Now I will go. I feel truly happy. . . . I am recalling how every morning when I left to go to elementary school I would say "See you later". Well, now I am leaving and I shall not return. I am truly satisfied' (p. 32), whilst a father writes to his daughters: 'Your father can't be the horse you ride on through life, so the two of you take care of each other. Please know your father is a happy man. He is riding a vehicle that will chase away our enemies. Become as great as your father' (p. 63). As such, these selected translations emphasize the extent of the adoption of the Myth of the War Experience discussed earlier and directly contradict accounts of the unhappiness with which many pilots contemplated their fate.

The theme of willing sacrifice is further exemplified in stories highlighted during the audio tour. One, also related elsewhere (Axell & Kase, 2002), tells of how a wife killed herself and her children so that her husband would not be held back from fulfilling his duty; another, based on a photograph of five young pilots aged between 17 and 19 who are smiling and playing with a puppy, suggests that they were happy even though they were allegedly departing on their mission the day after the photograph was taken. Nevertheless, it is also acknowledged that, in contrast, at night time many young pilots could be heard 'weeping bitterly . . . because they were afraid of death and felt a deep sorrow for their very short life' (Chiran, 2017: 19).

Overall, then, not only does the museum portray the young Japanese pilots as willing actors in the kamikaze campaign but also the narrative surrounding them emphasizes their individuality, youth and humanity, their honour and sacrifice. In other words, there is, as Allen and Sakamoto (2013: 1050) put it, a 'quasi hero worship . . . evident throughout the exhibits'. Not only is this focus on willing sacrifice 'consistent with Japanese revisionist approaches to the memories of the Asia-Pacific War' (Inuzuka, 2016: 157) but it is also a clear manifestation of dissonance as considered earlier in this paper. That is, the narrative presented at the museum can itself be challenged – and, indeed, has been through the stories of those pilots and others who survived the war (Sheftall, 2005) and who have, there-fore, been disinherited by the museum's interpretation – yet, of equal if not greater significance, other stories are also not told. No attempt is made, for example, to locate the kamikaze phenomenon within the context of the war or Japan's role in it; as one review on Tripadvisor (2020) notes, 'the museum merely glorified the sacri-fices made by these kamikaze pilots, and totally side-stepped the issues of Japanese aggression and war responsibilities. In my opinion, this made the pilots' deaths totally worthless'. No reference is made to the dissent of some commanders and pilots opposed to the kamikaze strategy or to its quite evident futility. Similarly, no reference is made to the prevailing culture of honourable death within the Japanese military at that time, nor the lure of promised deification; nor is it acknowledged that, in essence, those pilots were state-trained suicide bombers (Allen & Saka-moto, 2013). Moreover, also absent is any reference to their targets or victims, to the equally futile deaths of the crews of American ships that suffered direct hits.

   In short, the museum does not offer the knowledge or stories to enable its visi-tors, whether domestic or international, to begin to understand how and why, in the context of the last year of the Pacific War, the kamikaze phenomenon occurred and, in so doing, confront a difficult past. Rather, they are presented with a very specific, highly politicized and emotionally laden story that, for some visitors, might elicit feelings of respect and sympathy for the pilots but for others, bemusement and perhaps even anger. In turn, this raises the question of whether the Chiran Peace Museum, as a site of difficult heritage, might promote peace and reconciliation amongst its visitors, whether it is in fact 'peace museum'. On the one hand, many comments made by visitors, reproduced in a volume available for visitors to read in a seating area in the museum, suggest that it does. Typically, reference is made to the futility and waste of war and to the need for peace, although such arguably inevitable generic sentiments are largely expressed through the lens of the sacrifice of the young kamikaze pilots; that is, it is not the (untold) stories or heritages of the Pacific War that encourage visitors to write 'never again', but the senseless loss of thousands of young lives. Moreover, some visitors are of the opinion that the pilots' sacrifice laid the foundations of contemporary peace, a subliminal message that, perhaps, the museum intends to convey.

   On the other hand, other comments reveal a more critical perspective; one, for example, states: 'I am from Singapore . . . my country men also suffered . . . we lived through 3 half [*sic*] years of Japanese occupation', whilst another vis-itor (from Norway) simply observes: 'What a strange thing to call this place a peace museum!' (Chiran Visitors Book: 10–11). The latter comment reflects the conclusions of others who critique Chiran's interpretation of the kamikaze phe-nomenon, there being a broad consensus that a museum that not only honours the pilots as heroes but implies that their actions led to peace (whereas in fact they prolonged conflict) cannot be thought of as a peace museum. As Inuzuka (2016: 126) observes, peace 'is not integrated rationally into the displays' messages and pacifism is promoted in a rather ambiguous way'. Thus, whilst it may feed a con-temporary national identity amongst some Japanese visitors and, as a public insti-tution, reflects the national ambition for promoting peace through its name, the museum's quite evident dissonance, or its disinheritance of many potential tourists from both Japan and overseas, suggests that rather than stimulating understanding and reconciliation, it in fact denies (particularly domestic) visitors the opportunity to confront collectively and openly a difficult, controversial past.

9.4.2 *Yūshūkan War Museum*

Located within the grounds of (and managed by) Yasukuni Jinja, Yūshūkan is a highly controversial museum – indeed, to some, it is more problematic than Yas-ukuni (Fallows, 2014), yet, to date, it has attracted far more limited academic attention (e.g. Lambert, 2004; Yamane, 2009). It originally opened in 1882 to house artefacts from the Meij era Imperial Japanese Army, but the collection was expanded following the first Sino-Japanese War (1894–5). The building was demolished after an earthquake in 1923; following reconstruction, it opened again in 1932 and was subsequently expanded to include an interactive area that proved to be popular amongst visitors. According to Yoshida (2007), arguably reflecting public support for the country's militarism, 1.9 million people visited the museum complex in 1940. Following the Pacific War the museum was closed down and only opened again to the public in 1986 with a limited display, attracting rela-tively few visitors. However, after renovation and expansion, it reopened again in 2002 and 226,000 people visited the museum between July 2002 and May 2003 (Yoshida, 2007). Current visitor numbers and the balance between domestic and international visitors are not known, although a significant number of international tourists were observed in both Yasukini and Yūshūkan during the author's visit.

As a war museum, Yūshūkan is not dedicated specifically to the kamikaze although not only is there a memorial statue of a kamikaze pilot located near its entrance but also, as discussed shortly, the kamikaze feature prominently in one section of the museum, referred to in the English language leaflet as the 'Noble Spirits Sentiment Zone'. Rather, the museum presents an extensive collection of military artefacts dating back to the Meiji era but with a particular emphasis on twentieth-century conflicts, specifically the Asia Pacific War from 1937 onwards. Notably, five exhibition rooms and the so-called Great Exhibition Hall focus on the Pacific War, the latter including large exhibits such as a dive-bomber plane, a glider-bomber and a *kaiten* human torpedo. Interestingly, a model of a suicide diver is also on display in one of the smaller exhibition halls.

These physical exhibits are, as at Chrian, uncontroversial. Since 2002, how-ever, the focus of the museum has also been on education; as Takenaka (2015: 173) observes, the objective of the museum is 'educating the public on Japanese military history during the modern period and memorializing and honouring the spirits enshrined'. Consequently, much of the museum is given over to displays, many summarized in English, that provide a narrative of Japan's military activi-ties. And it is this narrative, conveying a highly revisionist history, that reflects the ownership and management of Yūshūkan by Yasukini Jinja that is so controversial. Inuzuka and Fuchs (2014: 31), for example, argue that the museum 'promotes a position of militarism disguised as self-defense' that is emphasized in a film with

English subtitles that is shown continuously. It was, according to the narrative, the United States that triggered the Pacific War, whilst the aggressors were European nations who had colonized those countries on whose resources Japan depended. More bluntly, Fallows (2014) writes:

> The museum is shocking in its mendacity. . . . It is entirely different to create a memorial to pay somber respect to those who died in a war . . . than it is to create a memorial that recasts an entire war in a glorified light, including over the widely recognized atrocities committed in that war.

For example, on entering Yūshūkan, the visitor is immediately confronted with a locomotive used on the infamous death railway between Thailand and Burma (Figure 9.3), yet reference is made neither to how and why that railway was built nor to the significant loss of life involved in its construction. Similarly, the widely acknowledged Nanking Massacre is referred to as an 'incident' with no mention of the atrocities committed by the Japanese.

Figure 9.3 Death Railway's locomotive, Yushukan War Museum.
Source: photo by Richard Sharpley

It is within this revisionist historical context promoting an imperialist ideology that the kamikaze are commemorated at Yūshūkan. Within the 'Noble Spirits Sentiments Zone' stands a statue of a kamikaze pilot (Figure 9.4), whilst cabinets display artefacts related to the kamikaze campaign. The walls of the rooms are covered with photographs of the pilots and other war dead, the noble spirits enshrined at Yasukuni (Figure 9.5). However, as at Chiran Peace Museum, no attempt is made to explain how and why the young men were encouraged to volunteer or to acknowledge the waste of young lives. Rather, by association with the message conveyed by the museum as a whole, the kamikaze are presented and commemorated as heroic participants in a conflict in which Japan fought gloriously to defend herself against Western aggression. In other words, not only is the kamikze heritage at Yūshūkan saturated with dissonance; it is also located within an inaccurate, militarist narrative that is more likely to elicit anger and disbelief than understanding and a desire for reconciliation amongst international visitors.

Figure 9.4 Statue of kamikaze pilot, Yushukan War Museum.
Source: photo by Richard Sharpley

Figure 9.5 Photographs of war dead, Yushukan War Museum.
Source: photo by Richard Sharpley

9.5 Conclusion

The kamikaze strategy arguably remains not only one of the most controversial Japanese military campaigns of the Pacific war but also one of the more misunderstood; the perception of young men seemingly willing to sacrifice their lives on behalf of their family, country and emperor contradicts a more nuanced and complex reality. Moreover, from an international visitor perspective, the kamikaze phenomenon perhaps epitomizes the approach of the Japanese military to the Pacific War, to fight and die nobly to the end, whatever the cost. Thus, it would be logical to suggest that, given Japan's stated objective of seeking, through international tourism, to enhance mutual understanding and to fulfil its responsibility to international peace (MLIT, 2012), the country's kamikaze heritage offers a potentially powerful means of contributing to understanding and encouraging more harmonious international relations.

From the evidence presented in this paper, however, this opportunity has been avoided; in other words, the presentation of kamikaze heritage at the two sites considered in this paper competes with Japan's official proactive stance on promoting international peace and understanding. This reflects, in part, the fact that the myth of the military experience propagated by Yasukuni Jinja (and which enjoys some support in contemporary Japan) directly shapes the narrative of Japanese military history presented at the Yūshūkan War Museum in particular. However, it also should be noted that the presentation of Pacific War heritage more generally cannot be separated from contemporary Japanese politics which, on the one hand, have tended to support an apologist approach towards the country's Asian neighbours but, on the other hand, remain imbued with a sense of victimhood (Tsutsui, 2009). Moreover, according to Nakano (2016: 165), more recent years have witnessed the emergence of a new political elite 'often opposed to expressions of war guilt and contrition' and driving a new tide of nationalism, illiberalism and historical revisionism. Hence, the manner in which the kamikaze are commemorated as revealed in this paper has undoubtedly been shaped by a complex amalgam of contextual influences (Farmaki, 2017) and continues to be so. This, in turn, supports the more general argument that not only may what Ashworth and Hartmann (2005) consider to be inevitable dissonance at difficult heritage sites be enhanced by dominant political and cultural factors but also that, consequently, the achievement of understanding and reconciliation through tourism to such sites may be both complex and challenging.

This may not always be the case. Nevertheless, to return to the specific purpose of this paper, it is evident that the nature of the commemoration and interpretation of the kamikaze pilots at both Chiran and Yūshūkan, as far as it is accessible to international visitors in terms of translated information, raises more questions than it answers. Certainly, a visit to Chiran Peace Museum in particular will leave international visitors with the sense that the kamikaze pilots were brave and honourable young men who died for their country. Yet, they will be left with questions about the prevailing political and cultural system that left the pilots with no choice and why contemporary Japan appears unable to acknowledge and accept responsibility for its wartime aggression. Moreover, given the overtly nationalistic and militarist narrative that is consumed by contemporary generations of Japanese domestic visitors at both sites, the potential for the strengthening of international relations in the future may also be questioned.

References

Allen, M. and Sakamoto, R. (2013) War and peace: War memories and museums in Japan. *History Compass*, 11(12), 1047–1058.
Ashworth, G. and Hartmann, R. (2005) *Horror and Human Tragedy Revisited: The Management of Sites of Atrocities for Tourism*. New York: Cognizant Books.
Axell, A. (2002) The kamikaze mindset. *History Today*, 52(9), 3–4.
Axell, A. and Kase, H. (2002) *Kamikaze: Japan's Suicide Gods*. Harlow: Pearson Education.
Bar-Siman-Tov, Y. (2004) Introduction: Why reconciliation? In Y. Bar-Siman-Tov (Ed.), *From Conflict Resolution to Reconciliation*. Oxford: Oxford University Press, pp. 3–9.
BBC (2017) How Japan's youth see the kamikaze pilots of WW2. *BBC News*. Available at: https://www.bbc.co.uk/news/world-asia-39351262 (Accessed 17 January 2020).
Blanchard, L. and Higgins-Desboilles, F. (2013) *Peace Through Tourism: Promoting Human Security Through International Citizenship*. Abingdon: Routledge.
Breen, J. (2004) The dead and the living in the land of peace: A sociology of the Yasukuni shrine. *Mortality*, 9(1), 76–93.
Buckley-Zistel, S. and Schaefer, S. (2014) *Memorials in Times of Transition*. Cambridge: Intersentia Ltd.
Cano, L. and Mysyk, A. (2004) Cultural tourism, the state and the Day of the Dead. *Annals of Tourism Research*, 31(4), 879–898.
Chiran (2017) *The Mind of the Kamikaze*. Minamikyushu-shi: Chiran Peace Museum.
Costello, J. (1982) *The Pacific War 1941–1945*. New York: Harper Collins.

Crowe-Delaney, L. (2019) Japanese tourism in the late 20th and early 21st centuries. In T. O'Rourke and M. Koščak (Eds.), *Ethical and Responsible Tourism: Managing Sustainability on Local Tourism Destinations*. Abingdon: Earthscan/Routledge, pp. 163–182.

D'Amore, L. (1988) Tourism: A vital force for peace. *Tourism Management*, 9(2), 151–154.

Danielsen, S. (2007) Japanese war movies aim to rewrite history. *The Guardian*, 23 May. Available at: https://www.theguardian.com/film/filmblog/2007/may/23/spanclassfloa trightbrsmallaeg (Accessed 7 April 2020).

Fallows, J. (2014) Stop Talking About Yasukuni; the Real Problem Is Yūshūkan. *The Atlantic*, 2 January. Available at: https://www.theatlantic.com/international/archive/2014/01/stop-talking-about-yasukuni-the-real-problem-is-y-sh-kan/282757/ (Accessed 20 January 2020).

Farmaki, A. (2017) The tourism and peace nexus. *Tourism Management*, 59, 528–540.

Friedrich, M. Stone, P. and Rukesha, P. (2018) Dark tourism, difficult heritage and memorialisation: A case of the Rwandan Genocide. In P. Stone et al. (Eds.), *The Palgrave Handbook of Dark Tourism Studies*. London: Palgrave Macmillan, pp. 261–289.

Funck, C. and Cooper, M. (2015) *Japanese Tourism: Spaces, Places and Structures*. New York/Oxford: Berghahn Books.

Gurler, E. and Ozer, B. (2013) The effects of public memorials on social memory and urban identity. *Procedia: Social and Behavioral Sciences*, 82, 858–863.

Hall, C.M., Timothy, D. and Duval, D. (2004) Security and tourism: Towards a new understanding. *Journal of Travel & Tourism Marketing*, 15(2–3), 1–18.

Harrison, D. and Sharpley, R. (Eds.). (2017) *Mass Tourism in a Small World*. Wallingford: CABI.

Ienaga, S. (2008) *Pacific War, 1931–1945: A Critical Perspective on Japan's Role in World War II*. New York: Presidio Press.

Inuzuka, A. (2016) Memories of the Tokko: An analysis of the Chiran Peace Museum for kamikaze pilots. *Howard Journal of Communications*, 27(2), 145–166.

Inuzuka, A. and Fuchs, T. (2014) Memories of Japanese militarism: The Yasukuni Shrine as a commemorative site. *Journal of International Communication*, 20(1), 21–41.

Jeans, R. (2005) Victims or victimizers? Museums, textbooks, and the war debate in contemporary Japan. *The Journal of Military History*, 69(1), 149–195.

JNTO (2019a) *Japan Tourism Statistics: Trends in Visitor Arrivals to Japan*. Japan National Tourism Organisation. Available at: https://statistics.jnto.go.jp/en/graph/#graph -inbound-travelers-transition (Accessed 24 December 2019).

JNTO (2019b) *2018 Visitor Arrivals and Japanese Overseas Travelers*. Japan National Tourism Organisation. Available at: https://www.jnto.go.jp/jpn/statistics/data_info_listing/pdf/190116_monthly.pdf (Accessed 24 December 2019).

Kelly, I. and Nkabahona, A. (2010) Tourism and reconciliation. In O. Moufakkir, O. and I. Kelly (Eds), *Tourism, Progress and Peace*. Wallingford: CABI, pp. 228–241.

Konstantopoulos, G. (2007) *The Kamikaze Pilots and Their Image in World War II*. Bachelor Thesis, Mount Holyoke College, South Hadley, MA. Available at: https://ida.mtholyoke.edu/xmlui/bitstream/handle/10166/731/228.pdf?sequence=1 (Accessed 17 January 2020).

Lambert, R. (2004) The maintenance of imperial Shinto in postwar Japan as seen at Yasukuni Shrine and its Yushukan Museum. *Asia-Pacific Perspectives*, 4(1), 9–18.

Light, D. (2017) Progress in dark tourism and thanatourism research: An uneasy relationship with heritage tourism. *Tourism Management*, 61, 275–301.

Litvin, S. (1998) Tourism: The world's peace industry? *Journal of Travel Research*, 37(1), 63–66.

Logan, W. and Reeves, K. (Eds) (2008) *Places of Pain and Shame: Dealing with 'Difficult Heritage'*. Abingdon: Routledge.

Lollis, E. (2014) Peace as a destination: Peace tourism around the world. In C. Wohlmuther and W. Wintersteiner (Eds.), *International Handbook on Tourism and Peace*. Klagenfurt: Drava, pp. 294–309.

Milano, C., Cheer, J. and Novelli, M. (Eds.). (2019) *Overtourism: Excesses, Discontents and Measures in Travel and Tourism*. Wallingford: CABI.

MLIT (2012) *Tourism20 Nation Basic Plan*. Ministry of Land, Infrastructure and Transport. Available at: https://www.mlit.go.jp/common/000234920.pdf (Accessed 22 December 2019).

MLIT (2016) *Meeting of the Council for a Tourism Vision to Support the Future of Japan*. Ministry of Land, Infrastructure and Transport. Available at: https://www.mlit.go.jp/com mon/001172615.pdf (Accessed 22 December 2019).

Moufakkir, O. and Kelly, I. (2010) *Tourism, Progress and Peace*. Wallingford: CABI.

Nakano, K. (2016) Political dynamics of contemporary Japanese nationalism. In J. Kingston (Ed.), *Asian Nationalisms Reconsidered*. Abingdon: Routledge, pp. 160–171.

Nelson, J. (2003) Social memory as ritual practice: Commemorating spirits of the military dead at Yasukuni Shinto shrine. *The Journal of Asian Studies*, 62(2), 443–467.

NHHC (2019) The first Kaiten suicide Torpedo attack, 20 November 1944. *Naval History and Heritage Command*. Available at: https://www.history.navy.mil/about-us/leadership/director/directors-corner/h-grams/h-gram-039/h-039-4.html (Accessed 21 January 2020).

Nitobé, I. (1908) *Bushido: The Soul of Japan*. Ebook available at: https://www.gutenberg.org/files/12096/12096-h/12096-h.htm (Accessed 2 December 2018).

Okuyama, M. (2009) The Yasukuni shrine problem in the East Asian context: Religion and politics in modern Japan. *Politics and Religion Journal*, 3(2), 235–251.

Oros, A. (2015) International and domestic challenges to Japan's postwar security identity: 'Norm constructivism' and Japan's new 'proactive pacifism'. *The Pacific Review*, 28(1), 139–160.

Orr, J. (2001) *The Victim as Hero: Ideologies of Peace and National Identity in Postwar Japan*. Honolulu: University of Hawai'i Press.

Palmer, C. (1999) Tourism and the symbols of identity. *Tourism Management*, 20(3), 313–321.

Pechlaner, H., Innerhofer, E. and Erschbamer, G. (Eds.). (2020) *Overtourism: Tourism Management and Solutions*. Abingdon: Routledge.

Pratt, S. and Liu, A. (2016) Does tourism really lead to peace? A global view. *International Journal of Tourism Research*, 18(1), 82–90.

Pye, M. (2003) Religion and conflict in Japan with special reference to Shinto and Yasukuni Shrine. *Diogenes*, 50(3), 45–59.

Richter, L. (1983) Tourism, politics and political science: A case of not so benign neglect. *Annals of Tourism Research*, 10(3), 313–335.

Rifai, T. (2013) Foreword. In C. Wohlmuther and W. Wintersteiner (Eds.), *International Handbook on Tourism and Peace*. Klagenfurt: Drava, p. 11.

Ryu, Y. (2007) The Yasukuni controversy: Divergent perspectives from the Japanese political elite. *Asian Survey*, 47(5), 705–726.

Sakamoto, R. (2015) Mobilizing effect for collective war memory. *Cultural Studies,* 29(2), 158–184.

Schäfer, S. (2016) From Geisha girls to the Atomic Bomb Dome: Dark tourism and the formation of Hiroshima memory. *Tourist Studies*, 16(4), 351–366.

Sharpley, R. (2009) Dark tourism and political ideology: Towards a governance model. In R. Sharpley and P. Stone (Eds.), *The Darker Side of Travel: The Theory and Practice of Dark Tourism*. Bristol: Channel View Publications, pp. 145–163.

Sharpley, R. and Friedrich, M. (2016) Genocide tourism in Rwanda: Contesting the concept of the 'dark tourist'. In G. Hooper and J. Lennon (Eds.), *Dark Tourism: Practice and Interpretation*. Abingdon: Routledge, pp. 134–146.

Sheftall, M. (2005). Japanese war veterans and kamikaze memorialization: A case study of defeat remembrance as revitalization movement. In J. Macleod (Ed.), *Defeat and Memory: Cultural Histories of Military Defeat in the Modern Era*. Houndmills: Palgrave Macmillan, pp. 154–174.

Sheftall, M. (2008) Japanese war veterans and kamikaze memorialization: A case study of defeat remembrance as revitalization movement. In J. Macleod (Ed.), *Defeat and Memory: Cultural Histories of Military Defeat in the Modern Era*. Houndmills: Palgrave Macmillan, pp. 154–174.

Shibuichi, D. (2005) The Yasukuni Shrine dispute and the politics of identity in Japan: Why all the fuss? *Asian Survey*, 45(2), 197–215.

Siegenthaler, P. (2002) Hiroshima and Nagasaki in Japanese guidebooks. *Annals of Tourism Research*, 29(4), 1111–1137.

Smith, L. (2006) *Uses of Heritage*. Abingdon: Routledge.

Takenaka, A. (2015) *Yasukuni Shrine: History, Memory and Japan's Unending Postwar*. Honolulu: University of Hawai'i Press.

Tripadvisor (2020) *Chiran Peace Museum*. Available at: https://www.tripadvisor.co.uk/Attraction_Review-g1022938-d1545999-Reviews-or20-Chiran_Peace_Museum-Minamikyushu_Kagoshima_Prefecture_Kyushu.html (Accessed 22 January 2020).

Tsutsui, K. (2009) The trajectory of perpetrators' trauma: Mnemonic politics around the Asia-pacific war in Japan. *Social Forces*, 87(3), 1389–1422.

Tunbridge, J. and Ashworth, G. (1996) *Dissonant Heritage: The Management of the Past as a Resource in Conflict*. Chichester: John Wiley & Sons.

Wight, A. C. (2016) Lithuanian genocide heritage as discursive formation. *Annals of Tourism Research*, 59, 60–78.

Williams, P. (2007) *Memorial Museums: The Global Rush to Commemorate Atrocities*. Oxford: Berg.

WTO (1980) *Manila Declaration on World Tourism*. Madrid: World Tourism Organization.

Yamamoto, T. (2002) *Bushido: The Way of the Samurai* (Based on the *Hagakure*: Trans. M. Tanaka, Ed. J. Stone). Garden City Park, NY: Square One Publishers.

Yamane, K. (2009) Moving beyond the war memorial museum. *Peace Forum*, 24(34), 75–84.

Yasuaki, O. (2002) Japanese war guilt and postwar responsibilities of Japan. *Berkeley Journal of International Law*, 20, 600–620.

Yoshida, K., Bui, H. and Lee, T. (2016) Does tourism illuminate the darkness of Hiroshima and Nagasaki? *Journal of Destination Marketing & Management*, 5, 333–340.

Yoshida, T. (2004) Whom should we remember? *Journal of Museum Education*, 29, 2–3.

Yoshida, T. (2007) Revisiting the past: Complicating the future. The Yushukan War Museum in Modern Japanese History. *History Faculty Publications*, Paper 2. Western Michigan University. Available at: http://scholarworks.wmich.edu/history_pubs/2 (Accessed 23 January 2020).

10 Chinese students confront the Hiroshima Peace Memorial Museum

Hamilton Bean

Dedicated to the participants in the Japan and the Pacific Century course, with hope for peace and reconciliation in East Asia.

In July 2018, I co-led a group of nine Chinese students to Japan for a two-week travel study course, "Japan and the Pacific Century," to explore the memorialization of the Pacific War, 1941–1945, and the role of public memory in East Asian affairs. For the Chinese, of course, the Pacific War begins much earlier, with Japan's invasion of China in 1931. Chinese typically draw moral equivalence between the atomic bombings of Hiroshima/Nagasaki in 1945 and the Japanese Army's massacre in Nanjing in 1938, as well as the inhumane experiments of Japan's Unit 731 in Manchuria (Kou, 2019). How these nine Chinese students confronted the public memory of the Pacific War during their visit to Hiroshima's Peace Memorial Museum – and what we can learn about peace and reconciliation as a result – is the focus of this essay. The essay responds to calls to examine whether heritage tourism involving sites with a controversial history can contribute to a better understanding among citizens of countries involved in war, whether this understanding can lead to lasting peace.

There is no "monolithic Japanese view" of the Pacific War (Jeans, 2005, p. 149), and the cross-disciplinary research literature concerning Hiroshima's place in Japan's public memory is vast (e.g., Hogan, 1996; Yoneyama, 1999; Zwigenberg, 2014). However, inclusion of the voices of actual visitors to the Peace Memorial Museum remains rare (Chen, 2012). This essay begins with background about the course, followed by a discussion of the course's broader economic and political context. The essay leverages scholarship concerning the apology from Rie Shibata (2018) and Keith Hearit (2006) to interpret the students' written reflections of their visit to the Peace Memorial Museum, providing a more granular understanding of why the Museum does not – and likely cannot – adequately promote reconciliation in East Asia.

10.1 Background about the course

The "Pacific Century" describes a 21st-century-dominated economically, politically, and culturally by the countries of the Asia-Pacific region, especially China, Japan,

DOI: 10.4324/9780367823795-14

and the United States. A conversation that I had in 2012 marks the origin of the Japan and the Pacific Century course. Specifically, I undertook a teaching assignment that year at the University of Colorado Denver's International College Beijing (ICB). Instructors would occasionally gather in the evenings on the rooftop of Building 41 (a large apartment complex on campus) to admire the city view and talk about the joys and challenges of teaching in China. One night on the rooftop, a fellow instructor relayed a teaching experience that he had had with a 12-year-old Chinese girl. He had done an activity in class that day where he drew a picture of the Earth on the whiteboard with an asteroid headed towards it, explaining to the students,

You've just intercepted a secret transmission from NASA and found out there is a huge asteroid heading for Earth. We're all doomed and only have 24 hours to live, but only you know about it. My question is: How are you going to spend these last hours on Earth?

The girl, a diligent student who seldom spoke in class, calmly replied, "I'd take all the planes I could find and fly to Japan and bomb all the Japanese." The girl's response to the question was not uncommon; my colleague later conducted the same activity in high school classes and received similar answers. When he tried to point out to these students that, given the circumstances, it was pointless to kill all the Japanese, the students were not swayed – they thought it was worthwhile bombing the Japanese anyway. Whether or not these students had been taught that it was desirable to perform such extreme displays of "patriotic education" for their classmates, the comments seemed to reveal an intense hatred of the Japanese. These events occurred in 2012 during a spike in anti-Japanese rhetoric and propaganda following the Diaoyu/Senkaku Islands dispute; therefore, the remarks may have been more reflective of official government position rather than students' genuine attitudes.

Nevertheless, the story was stunning: I resolved to learn more about China-Japan relations with the aim of someday taking Chinese students to Japan so that they might gain a firsthand view of the country and (I hoped) reconsider their patriotic education and taken-for-granted assumptions about the Japanese. I had taught English at the AEON language school in Wakayama, Japan, from October 1996 to December 1997, and I held the Japanese people in positive regard (while not always agreeing with the Japanese government's policies). As an admirer of Ezra Vogel's (2019) scholarship (Vogel being one of the few U.S. scholars to maintain deep knowledge of and connections to both China and Japan), I was eager to build bridges among China, Japan, and the United States. However, it would take six more years before I was finally able to organize and conduct the envisioned course. While most of those six years saw consistent China-Japan animosity, the U.S.-China trade war that began in 2018 (on the eve of our course) ushered in improved ties between Beijing and Tokyo. America's so-called "unsinkable aircraft carrier in the Pacific" (as Japan had been called during the Cold War; McGregor, 2018, p. 13) was rapidly exploring new levels of cooperation with China.

Figure 10.1 Japan and the Pacific Century Course students and instructors.

Over the fall and spring of 2018, I developed the "Japan and the Pacific Century" course, as well as arranged for another communication instructor and ICB colleague to co-lead it. My then-13-year-old son accompanied us as well. On July 16, 2019, the three of us, along with the nine Chinese college students (five males and four females), arrived in Japan to begin our course (see Figure 10.1). The situation was unusual: two American professors leading nine Chinese students through Japan to confront sites of public memory of the Pacific War. All of us were aware of the peculiarity of our course, but it can be understood in the context of Asia's "war memory boom," which Frost, Vickers, and Schumacher (2019) describe as part of "a wider process of Asian heritage-making, whose focus has increasingly shifted from the ancient past to incorporate modern, especially twentieth century, sites and remnants" (p. 2). Our course was symbolic of how the "explosive emergence" of war memory tourism has suddenly become a fixture in East Asia (p. 20).

The course ran until July 29, 2018, and we traveled through sites in Osaka, Nara, Koyasan, Kyoto, Hiroshima, and Miyajima. We focused on Pacific War-related museums, memorials, and landmarks that engaged Japanese public memory and national identity, e.g., Peace Osaka, Osaka Museum of History, Koyasan, Kyoto Museum for World Peace, Hiroshima Peace Memorial Museum, and others. Our learning objectives were to: (a) understand and articulate key concepts related to the construction, maintenance, and transformation of narratives about Japan, the Pacific War, and contemporary Pacific affairs; (b) conduct analysis of official statements, policies, media, and public texts related to the Pacific War in order to critically assess their meanings and implications for various audiences including scholars, officials, citizens, and international publics; and (c) write about the influence of Pacific War-related sites, displays, and events in Japan from a critical perspective. How the Hiroshima Peace Memorial Museum related to one of our course concepts – reconciliation – is discussed next.

10.2 Hiroshima and reconciliation

Tourism in Japan has risen dramatically in the last six years, with each year producing a record level of visitors (Eiraku, 2019). In 2018, China accounted for more than eight million visitors to Japan, with China being the single largest country of origin of foreign tourists. The influx is growing (although the 2020 Covid-19 pandemic will interrupt this trend): 2018 saw a record high, and a nearly 14 percent increase, in Chinese visitors from the year before (2019 figures were unavailable at the time that this chapter went to press). Despite the major growth in tourism, a visit to the Hiroshima Peace Memorial Museum is not on the itinerary for most Chinese visitors. According to the Japan Tourism Agency, Hiroshima was not among Chinese tourists' top 16 most visited prefectures in 2018 (2019 figures were unavailable; the 2018 survey reported only the top 16 prefectures). Chinese visitors accounted for less than 10 percent of the overall number of foreign visitors to Hiroshima that year (Nippon.com, 2019). Increased Chinese tourism in Japan, and simultaneous avoidance of Hiroshima, formed the backdrop of our course as it commenced in July 2018.

Chinese attraction/ambivalence towards Japan reflects broader reconfigurations of power in East Asia. Richard McGregor's (2018) *Asia's Reckoning: China, Japan, and the Fate of U.S. Power in the Pacific Century* identified the main drivers of these reconfigurations. McGregor argues that decades' worth of close economic ties between China and Japan have failed to produce strong political bonds, with periodic downward spirals marking the last decade or so in particular. A primary contributor to the mutual ill-will, according to McGregor, is the legacy of the Pacific War, with Chinese and Japanese interlocutors continuing to battle over public memory of the nearly 75-year-old conflict. Today, the battlelines are found in public education curricula, books, magazines, movies, documentaries, and in academic fora, among other sites. For McGregor, "China in particular has a whiff of the Balkans, where many young people have a way of vividly remembering wars they never actually experienced" (p. xv). Yet, when considering Japan's brutal invasion and occupation of China, it is easy to understand why. In 2016, I accompanied a group of Chinese students to the "Memorial Hall of the Victims in the Nanjing Massacre by Japanese Invaders." After our visit, as we waited for the bus to return to our hotel, I observed that the experience of the memorial had muted the normally cheerful and talkative students. One of the most mild-mannered of the students finally declared: "I hate the fucking Japanese!"

McGregor's observations concerning China's patriotic education help to account for the high school students' responses to my colleague's question that initially sparked the idea for the Pacific Century course. But Japan has not been immune to the same nationalistic impulses. When I returned to Japan in 2018, I was dismayed that right-wing revisionism had grown in the 22 years since I had lived in the country. A rightward lurch in film, television, and public attitudes had eroded the 1990s trend of better acknowledging Japan's colonialism and wartime aggression (Nakano, 2015). But even in the 1990s, the rightward shift was already on the horizon. Specifically, a plan to include an exhibition of Hiroshima's role in Japan's colonial past within the Peace Memorial Museum had, by 1994, been significantly scaled back to fit with the vague "Spirit of Hiroshima" theme (Jeans, 2005; Naono, 2005). Although I had read of Japan's rightward trajectory throughout the 2000s, seeing the evidence of it with my own eyes during our course was jarring (but a broader discussion of this trend lies outside the scope of this essay).

Shibata (2018) argues, "How the transgressor and transgressed deal with their past history of violent trauma is a critical component to reconciliation in East Asia" (p. 294). Shibata's claim served as a touchstone throughout our course, and students explicitly engaged it in their final reflection assignment. Our course included more than a dozen readings, but Shibata's chapter, "Apology and Forgiveness in East Asia," played a critical role. Shibata claims that reconciliation in East Asia hinges on actions that address the deep emotional and psychological needs of both victims (i.e., needs for recognition and restoration of self-esteem) and perpetrators (i.e., needs for protecting one's identity as a moral actor). For Shibata, the Japanese government's attempts to redress past injustices have defended Japan's identity as a moral actor but failed "to satisfy the fundamental needs for recognition and restoration of self-esteem amongst the victimized nations" (p. 273). Complementing Shibata's chapter, John Dower's (2014) *Ways of Forgetting, Ways of Remembering: Japan in the Modern World* provided a useful typology of war memory in Japan: denial, moral equivalence, victim consciousness, sanitization, and acknowledgment of guilt and responsibility. These categories became the lenses through which the students in the course interpreted the displays of public memory that we confronted.

The short timeframe of the course prevented a more in-depth reading of the vast literature concerning the struggle over Hiroshima's place in Japan's public memory. Several studies would have been particularly useful for our purposes. Specifically, Benedict Giamo's (2003) "The Myth of the Vanquished: The Hiroshima Peace Memorial Museum" provides an insightful analysis of the Museum's self-serving omissions. Giamo is blunt: "the Hiroshima Peace Memorial Museum and Park does not wade into forbidden territory. Context and conjecture are not simply foreshortened, they are deliberately suppressed" (p. 704). Roger Jeans' (2005) "Victims or Victimizers? Museums, Textbooks, and the War Debate in Contemporary Japan" describes the short-lived push for an "aggressors' corner" in the Museum in the late 1980s. Understanding that history would have helped the Chinese students appreciate the museum as a site of struggle. Likewise, Akiko Naono's (2005) analysis of a decade-long conflict to better acknowledge Japan's colonial past within the Museum highlights the difficulties of addressing the omissions that Giamo and Jean describe. Chia-Li Chen's (2012) analysis of the Museum's visitor comment books highlights the diversity of responses to the exhibits, while Thomas Olesen (2019) accounts for this diversity in unpacking the dynamics of Hiroshima as an international "memory complex." Recent work by Kazuyo Yamane (2017) and Jooyoun Lee (2018) explains both the enduring and changing roles of war and peace museums in Japan.

It would also have been useful for the students to have engaged in scholarship concerning institutional apology in order to gain a more granular understanding of how apology is performed rhetorically. Specifically, for Hearit (2006), institutions accused of wrongdoing, such as governments, often need to perform symbolic and ritualistic acts of mortification to restore institutional identity, public values, and social harmony. However, few studies have focused on the substance, style, and situations that adhere to institutional apologies. Extant studies suggest that there are five strategies that institutions rely upon in their apologetic discourse: confession or mortification, corrective action, compensation, transcendence (e.g., shifting blame for the Pacific War to the Western powers that Japan had fought on behalf of Asian nations), and ritual (e.g., ceremonies, pageants, and official events). Hearit and others have argued that an ethical apology explicitly acknowledges wrongdoing, fully accepts responsibility, expresses regret, identifies with injured stakeholders, asks for forgiveness, seeks reconciliation with injured stakeholders, fully discloses information related to the offense, provides an explanation that addresses legitimate expectations of the stakeholders, offers to perform an appropriate corrective action, and offers appropriate compensation. Understanding institutional apology in this way would have complemented Shibata's (2018) analysis of the Japanese government's public apologies (and their negations), as well as have made students sensitive to the difficulty (or perhaps impossibility) of performing adequate institutional apology within the constraints of museum exhibits.

Scholarship concerning Hiroshima's place in Japan's public memory has already extensively documented the Peace Memorial Museum's self-serving omissions. This essay's contribution therefore lies in amplifying the voices of Chinese visitors, which appear to be mostly absent from studies of the site. For example, the relative infrequency of Chinese visitors to the Museum, among other factors, prevented Chen (2012) from pinpointing that group's reactions in her analysis of visitor comment books. Extant studies typically rely on ethnographic approaches that privilege the authors' interpretations and experiences. Even studies that include the perspectives of visitors tend to rely on generalized summaries rather than written reflections. The essay next turns to the Chinese students' reflections.

10.3 Confronting Hiroshima

The students' reflections must be considered in light of the extensive renovations of the Hiroshima Peace Memorial Museum that occurred between 2016 and 2019. The museum's Main Building was closed during our course. Visitors were instead directed to the newly renovated East Building, which temporarily housed some of the Main Building's exhibits, artifacts, and displays. The Main Building reopened in 2019; therefore, the students' assignments reflected a layout of the Museum that no longer exists. It can nevertheless be assumed that certain exhibits, artifacts, and displays are similar or identical between the museum's 2018 and post-renovation versions. According to a renovation information flyer that we received, the Main Building houses *The Reality of the Atomic Bombing* exhibit, which conveys the inhumanity of the atomic bomb, the severity and atrocity of the atomic bombing, and the anguish and sorrow of the victims and their families. The East Building houses *The Dangers of Nuclear Weapons* exhibit and the *Hiroshima's Progress* exhibit. The former conveys the events leading up to the bombing and the concern that the existence of nuclear weapons is a threat to all humanity, while the latter introduces the city's progress from the time of the war, through its recovery, to its effort to become a city of peace. While our group could only access the East Building, it was clear that material from the Main Building's *The Reality of the Atomic Bombing* exhibit had been moved to the temporary East Building location.

Nevertheless, it is important to note that there may have been displays of the Nanjing Massacre that were unavailable for the students to see during our visit; specifically, one that included a photograph of Hiroshima residents celebrating Nanjing's fall with a lantern parade. According to Jeans (2005), this photo included the caption: "Hiroshima's citizens celebrating with a torchlight parade. In Nanjing, however, Chinese were being massacred by the Japanese Army" (p. 170). Many scholars and commentators have noted that the Hiroshima Peace Memorial Museum pays comparatively little attention to Japan's invasion and occupation of China (e.g., Lee, 2018), and for some students, the museum's overwhelming focus on the Japanese victims of the atomic bombing generated sympathy but failed to spur any feelings of forgiveness or reconciliation. Upon entering the museum, the students immediately sensed what Dower (2014) described as "victim consciousness." One student wrote, "We visited Hiroshima Peace Memorial Museum. I felt the shock of strong victim consciousness there. There are a large number of sensational pictures of victims in there." Another student observed, "the exhibition just tells us the part of the story. It is select[ive] and emphatic, it mainly focuses on the victim side, shaping Japan is a sufferer of the war." This student concluded, "The victim consciousness traps Japan and prevents the country from facing the history and the atrocit[ies] they did." Another student wrote,

> From my perspective, both of the museum and memorial told people the damage of the nuclear weapons and the condition of the people who live in Hiroshima at that time. However, they did not tell people the reason[s] why America [had to] bomb . . . Japan and their [Japan's] behaviors in China. In my own opinion, it is a kind of behavior that they avoid to describe the truth to the public.

For one student, the absence of depictions of China's suffering during the Pacific War nullified Japan's bid as victim: "The Chinese suffered too much damage from

Japan in World War II. So, the Chinese cannot admit that Japan is also a victim. From the Chinese point of view, [the] Japanese are the sponsor of the war, not the victims." Another student asked the question, "Is it appropriate to emphasize domestic harm and ignore Japan's harm to other countries?" This student's answer was clear:

> Museums [in] Japan lack images of the countr[ies] Japan invaded. The dignity of the Japanese forces them to choose a selfish action, forcing them to cover [over] Japanese aggression by promoting the idea of peace. However, this double standard has become the "trigger" for the untreatable postwar trauma. Covering up the facts is an irresponsible approach to history and will slow down the reconciliation process.

This student also noticed that the museum's inadequacy vis-à-vis Chinese war memory could be productively contrasted with the Japan-U.S. relationship:

> As . . . Dower said, Japan's hatred against the American[s] ha[s] never become a dominant sentiment. . . . Japanese quietly mourn the atomic bomb victims in the memorial hall with victim consciousness, but without any hatred [toward Americans]. I think there are two explanations [for] it. First of all, both Japanese and Americans can acknowledge that Japanese are "both perpetrators and victims". . . . The acknowledgement [is] a kind of . . . war memory [that] benefits the reconciliation between transgressor and transgressed. As I think more deeply, I think it has something to do with the U.S.-Japan sanitization. This means that Japan and the United States are simultaneously trying to avoid certain histories.

This student's insight helps to explain the group's criticism of the museum: both Japan and the United States have long "sanitized" each other's roles as war perpetrators in order to facilitate the post-war security alliance. In the museum, sanitization is apparent in the perfunctory discussion of Japanese wartime aggression (more on this below) and the lack of reconsideration of questionable U.S. tactics. Regarding those tactics, as former U.S. Secretary of Defense Robert McNamara recalled of U.S. Air Force General Curtis LeMay's civilian bombing campaign in Japan in the film *The Fog of War*:

> LeMay said, "If we'd lost the war, we'd all have been prosecuted as war criminals." And I think he's right. He, and I'd say I, were behaving as war criminals. LeMay recognized that what he was doing would be thought immoral if his side had lost. But what makes it immoral if you lose and not immoral if you win?
>
> (Morris, 2005)

The answer to McNamara's question relies, in part, on how the winners and losers memorialize the conflict and the factors that drive that decision. As Olesen

(2019) explains, "In return for the Japanese not portraying Hiroshima as a victim of war crimes and injustice . . . the United States would not pursue direct condemnation of Japan's Asia policies" (p. 6). Since the high point in Japan-China bilateral relations from the 1970s to 1990s, the two countries have not had a strong alliance that has required the Chinese government to sanitize Japan's wartime atrocities. The Japanese government, in turn, has not, at least in recent years, had sufficient cause to placate China in the ways that it has historically done with the United States. As a result, the Hiroshima Peace Memorial Museum cannot be expected to address the psychological needs of Chinese visitors. As one student in the course summed up:

> After visiting Hiroshima, I am aware of how the transgressor and transgressed deal with their past history of violent trauma is a critical component to reconciliation in East Asia. . . . In my opinion, Japan's demand is to be recognized not only [as] a perpetrator, but also [as] a victim. The demands of the invaded countries of World War II were Japanese sincere apologies in language and action. It is obvious that relationships between other east Asian countries and Japan [are] not as good as that of America and Japan because Asian countries' needs of war memory cannot be satisfied.

While most students found the museum's overwhelming emphasis on victim consciousness lamentable, a few students noted the museum's acknowledgment of Japan's wartime transgression in Nanjing. Several students remarked about its significance:

> The visit to the Hiroshima Peace Memorial Museum made me feel a little relief because I saw that instead of just mention[ing] how terrible the atomic bomb was, which makes up most of the exhibition and expressed the sense of victim consciousness, the museum also mentions the Nanking massacre with the specific death toll on it as well, although just a few sentences makes me, as a Chinese, feel relief. I even excitedly told my partners that they [the museum] mentioned the Nanking massacre.

Another student wrote:

> The Hiroshima Peace Memorial Museum shows the historical event of the Nanjing Massacre and records the official number of victims, 300,000, which is the official number issued by the Chinese government. And the statement objectively states the historical facts of the fall of Nanjing. Even though it doesn't elaborate the violence of Japanese Army, it is still a sign of progress.

Another student remarked:

> In yesterday's tour, the objective attitude of the Nanking Massacre is a good example for both countries. In the Memorial Hall for the Victims,

it not only showed its respect to Japanese people who died in the atomic bombing but also to the victims who died in the Nanking Massacre, "The Japanese army attacked Nanjing . . . the number of victims exceeded 300,000". . . . In fact, it's the first time [we] see [a] Japanese official statement to admit the accurate number of victims instead of a denial. As a Chinese, I can accept its emotion and the message it sends, cherishing peace. I think the Japanese people are starting to wake up to that history, it is a good signal.

Yet, another student was dissatisfied with the display:

Although I also saw the mention about the Nanjing Massacre and the "misguided policies" in the museum, they only took a trivial place in the museum compared to those contents expressing Hiroshima as the victim of war, and the statement of those atrocities that the Japanese Army had done in China was not related to the overall situation during the war, hence people who were unfamiliar with this phase of history would not consider the possible causality between Nanjing and Hiroshima.

Hopeful signs aside, the insufficiency of the Hiroshima Peace Memorial Museum in meeting the students' emotional and psychological needs prompted some students to consider what might be done in the future to promote reconciliation. One student wrote,

The real apology is not how many statements the Japanese government has made in the official sense, but instead, [how] they show that they are aware of their mistakes. When they are genuinely willing to face history and face the past, problems can be solved completely.

Presumably, for this student, evidence of Japan's genuine willingness to face history would include more explicit and sustained acknowledgment of Japan's wartime aggression within the Hiroshima Peace Memorial Museum, as well as other sites in Japan. Another student echoed this sentiment:

[E]ven though the sites in Hiroshima are keen to appeal for peace in the future, all of them have a common defect, which is lacking investigating [Japan's] own responsibility for the war, just as Shibata mentioned that the perpetrator's acceptance of responsibility is one of the most effective elements to make an apology.

Another student wrote:

They [the Japanese] emphatically suggest the price they paid for the war instead of the cause of the war, which is not a thorough way to prevent the war in the future. To build a peaceful global community, personally I think people should mourn for not only the war but also [conduct] self-examination the origin of the war.

One student, however, perceived a self-serving motive for the Chinese government in criticizing Japan's lack of apology:

On the other side, China is also taking advantage of the transgressed identity for national interests. Personally, I think one of the reasons why China is emphasizing the past history is to strengthen the nationalism among Chinese people to consolidate the rule and show the voice and power in the world. Even though the past history and violent trauma are precisely the facts, I think the primary purpose of the Chinese government is using the past as a weapon to face the current challenges.

In sum, the students' reflections illustrate the Hiroshima Peace Memorial Museum's inability to meet the emotional and psychological needs of Chinese visitors. The museum does not aim to address these needs, and as one might expect, the exhibits fall far short of the characteristics of effective apology noted by Hearit (2006), Shibata (2018), and others. As Giamo (2003) concludes, "the vivid recollections and glaring omissions display a disturbing form of dissociation that defines the myth of the vanquished" (p. 703). Questions thus arise concerning whether the Peace Memorial Museum, or any museum in Japan, can meaningfully help counteract this myth. Although the students' comments reflect the entrenched "perpetrator/victim" binary, scholars have described how other sites in Japan better recognize "the complexity of how Japan's controversial wartime past is presented in the Japanese public domain" (Allen & Sakamoto, 2013, p. 1047). How can heritage tourism involving sites with a controversial history contribute to a better understanding among citizens of countries involved in war? Can this understanding lead to reconciliation and lasting peace? The essay next explores these questions.

10.4 Lingering questions

The students' reflections reveal how the Hiroshima Peace Memorial Museum's omissions and cursory gestures concerning its colonial past, wartime aggression, and post-war alliance with the United States mirror the Japanese government's inability to adequately reconcile with its East Asian neighbors. The museum does not adequately address Chinese visitors' emotional and psychological needs, nor does it aim to. It is therefore easy to understand why Chinese visitors in Japan tend to avoid the museum. In absence of a transformation in museum stakeholders' willingness and ability to critically confront Japan's past, Chinese visitors will likely continue to avoid Hiroshima. Matthew Allen and Rumi Sakamoto's (2013) study of six sites and museums in Japan (Yushukan, Chiran Peace Museum for Kamikaze Pilots, Showakan, Shokeikan Museum for Wounded Soldiers, Okinawa Prefectural Peace Park Memorial Museum, and the Himeyuri Memorial and Museum) demonstrated that these institutions use "peace" to frame the events of the Pacific War to allow "people's roles in these events to be acknowledged publicly and enabling these experiences to be rehabilitated in the public domain" (p. 1048). A logical next step for the current project, then, would be to assess how Chinese visitors make sense of other sites of Pacific War memory in Japan.

In hindsight, I believe that I was naïve in my assumption that simply visiting Japan would somehow soften Chinese students' critical attitudes about the Japanese. Our course potentially reinforced those attitudes. To see whether this was true, I contacted the students about one year after our course had concluded to re-survey them about their attitudes. About halfway through our course, I surveyed the students about their attitudes toward the Japanese using Pew Center (Stokes, 2016) survey questions. One question was, "Do you have a very favorable, somewhat favorable, somewhat unfavorable, or very unfavorable opinion of Japan?" At the time, two students reported a very favorable opinion, four a somewhat favorable opinion, three a somewhat unfavorable opinion, and one a very unfavorable opinion (the course's co-instructor participated in this survey). One year later, in preparation for writing this essay, I asked the students to respond to the same questions again. This time, three students reported a very favorable opinion, five a somewhat favorable opinion, and two a somewhat unfavorable opinion. These responses indicated a slight improvement in the students' attitudes toward Japan.

In the initial survey, I also followed the Pew Center in asking the students to associate various traits with the Japanese people. The students universally described the Japanese as "hardworking" and "modern." Seven students, however, also described the Japanese as "nationalistic." A year later, these numbers were unchanged, except that three students had demurred in describing Japan as "modern" (perhaps reflecting China's breathtaking advancements in AI, mobile payment systems, and infrastructure). I also asked, both during the course and a year later, whether the students thought that Japan had apologized sufficiently for its military actions during the 1930s and 1940s. Initially, one person thought that Japan had apologized sufficiently, while nine others said that Japan had not apologized sufficiently. One year later, only one more person agreed that Japan had apologized sufficiently. Both surveys were obviously non-scientific, but it is difficult to conclude that the course had much influence on students' attitudes over the one-year intervening time period.

Of course, people often perceive the influence of events in their lives only much later, but for stakeholders hoping to use travel study in Japan to promote China-Japan reconciliation (as I had hoped), the outcomes reported here are disheartening. These outcomes are unsurprising given that the students confronted repeated omissions and obfuscations of Japan's invasion and occupation of China. Even sites that ostensibly confronted Japan's colonial history in a critical way (such as Peace Osaka) left the students emotionally and psychologically deflated. Carol Gluck (On the Media, 2016) argues that ordinary Japanese citizens hold different views of history than government officials, but even when I attempted to introduce the Chinese students to my Japanese friends, interaction was limited.

Interpersonal interactions aside, the students' reflections concerning Hiroshima point to the difficulty of leveraging the Peace Memorial Museum, specifically, and Japan's war and peace museums, more generally, in promoting reconciliation. Scholars often point to Germany as a model for how to perform reconciliation, but in Japan, the difficulty exists at two levels. The first level is within the Peace Memorial Museum itself. As study after study has made clear, it is unlikely that museum stakeholders would permit the kinds of messages that could adequately meet Chinese visitors' emotional and psychological needs. Hearit (2006), Shibata (2018), and others have argued that an ethical apology explicitly acknowledges wrongdoing, fully accepts responsibility, expresses regret, identifies with injured stakeholders, asks for forgiveness, seeks reconciliation with injured stakeholders, fully discloses information related to the offense, provides an explanation that addresses legitimate expectations of the stakeholders, offers to perform an appropriate corrective action, and offers appropriate compensation. Even if museum stakeholders permitted a more thorough accounting of Japan's colonial past, wartime aggression, and U.S.-Japan sanitization, it is difficult to see how any exhibit, or set of exhibits, could adequately convey the content needed for an ethical and effective apology.

The second level of difficulty is that even if the Peace Memorial Museum were somehow able to communicate an ethical and effective apology, it would still remain only one location within a web of institutional discourse, both in Japan and internationally. The museum is only a single node within the broader Hiroshima "memory complex" that shapes visitors' experiences. As Olesen (2019) explains:

> Viewing Hiroshima as part of a memory complex is to disrupt any assumed singularity and homogeneity about it; to bring into focus how the same event is involved in a constant negotiation over both memory ownership and meaning. When Hiroshima is remembered through or with Pearl Harbor, Nanjing, and so on, that inevitably affects the interpretation of the Hiroshima memory itself, as well as vice versa.
>
> (p. 2)

Olesen's argument renders any generalizations about Hiroshima based on the students' reflections problematic. Yet, this case study suggests that confrontations with the Hiroshima Peace Memorial Museum, and sites like it, are unlikely to stimulate yearnings for reconciliation among Chinese visitors predisposed to view Japan as the perpetrator of violent aggression in East Asia. This conclusion may be pessimistic, but it is one based on the actual reflections of Chinese visitors, rather than conjectures about what visitors feel and believe. The conclusion may be irrelevant, however, in that the students in our course appeared ready to move on, believing that Japan is incapable of apologizing in ways that satisfy the needs of the transgressed. To illustrate this point, on the last day of our course, I intimated that I feared that China was becoming less interested in gaining Japan's apology and more interested instead in "payback" for past transgressions via assertions of

China's economic and military power. China increasingly appears to favor confrontation, not reconciliation (Pomfret, 2019). The students did not deny my speculation. I thus concluded then, and maintain now, that in absence of apology, the transgressed will seek other ways of meeting their emotional and psychological needs for recognition and self-esteem. In the context of China-Japan relations, we can hope that those needs may be met without risking further wars in East Asia, but history does not inspire confidence.

References

Allen, M., & Sakamoto, R. (2013). War and peace: War memories and museums in Japan. *History Compass*, *11–12*, 1047–1058.

Chen, C. L. (2012). Representing and interpreting traumatic history: A study of visitor comment books at the Hiroshima Peace Memorial Museum. *Museum Management and Curatorship*, *27*, 375–392.

Dower, J. W. (2014). *Ways of forgetting, ways of remembering: Japan in the modern world*. New York: The New Press.

Eiraku, M. (2019). *Foreign visitors to Japan hits record high*. NHK. https://www3.nhk.or.jp/nhkworld/en/news/backstories/347/

Frost, M. R., Vickers, E., & Schumacher, D. (2019). Locating Asia's war memory boom: A new temporal and geopolitical perspective. In M. R. Frost, D. Schumacher, & E. Vickers, *Remembering Asia's World War two* (pp. 1–24). London: Routledge.

Giamo, B. (2003). The myth of the vanquished: The Hiroshima Peace Memorial Museum. *American Quarterly*, *55*, 703–728.

Hearit, K. M. (2006). *Crisis management by apology: Corporate response to allegations of wrongdoing*. Mahwah, NJ: Lawrence Erlbaum Associates, Inc.

Hogan, M. J. (Ed.). (1996). *Hiroshima in history and memory*. Cambridge: Cambridge University Press.

Jeans, R. B. (2005). Victims or victimizers? Museums, textbooks, and the war debate in contemporary Japan. *The Journal of Military History*, *69*(1), 149–195.

Kou, K. (2019, August 5). Chinese opinions on Hiroshima and Nagasaki. *SupChina*. https://supchina.com/2019/08/05/kuora-chinese-opinions-on-hiroshima-and-nagasaki/

Lee, J. (2018). Yasukuni and Hiroshima in clash? War and peace museums in contemporary Japan. *Pacific Focus*, *33*, 5–33.

McGregor, R. (2018). *Asia's reckoning: China, Japan, and the fate of U.S. power in the Pacific century*. London: Penguin Books.

Morris, E. (2005). *The fog of war*. Culver City, CA: Sony Pictures Home Entertainment.

Nakano, K. (2015). The legacy of historical revisionism in Japan in the 2010s. In G. Rozman (Ed.), *Misunderstanding Asia: International relations theory and Asian studies over half a century* (pp. 213–225). New York: Palgrave Macmillan.

Naono, A. (2005). 'Hiroshima' as a contested memorial site: Analysis of the making of a new exhibit at the Hiroshima Peace Museum. *Hiroshima Journal of International Studies*, *11*, 229–244.

Nippon.com. (2019). *From snow country to Hiroshima: International tourist preferences*. https://www.nippon.com/en/japan-data/h00513/from-snow-country-to-hiroshima-international-tourist-preferences.html

Olesen, T. (2019). The Hiroshima memory complex. *The British Journal of Sociology*, *71*, 81–95.

On the Media. (2016, May 19). *Visiting Hiroshima without revisiting history*. https://www.wnycstudios.org/podcasts/otm/segments/obama-will-visit-hiroshima-he-wont-revisit-history

Pomfret, J. (2019, December 18). It's not all on Trump: China favors confrontation with the U.S. *Washington Post*. https://www.washingtonpost.com/opinions/2019/12/18/its-not-all-trump-china-favors-confrontation-with-us/

Shibata, R. (2018). Apology and forgiveness in East Asia. In K. P. Clements (Ed.), *Identity, trust, and reconciliation in East Asia* (pp. 271–297). Cham, Switzerland: Palgrave Macmillan.

Stokes, B. (2016). *Hostile neighbors: China vs. Japan*. Pew Center. https://www.pewresearch.org/global/2016/09/13/hostile-neighbors-china-vs-japan/

Vogel, E. F. (2019). *China and Japan: Facing history*. Cambridge, MA: Belknap Press.

Yamane, K. (2017). Japanese peace museums: Education and reconciliation. In A. Hunter (Ed.), *Peace studies in the Chinese century* (pp. 101–130). New York: Routledge.

Yoneyama, L. (1999). *Hiroshima traces: Time, space, and the dialectics of memory*. Berkeley: University of California Press.

Zwigenberg, R. (2014). *Hiroshima: The origins of global memory culture*. Cambridge: Cambridge University Press.

Conclusion

Rudi Hartmann

A multitude of changes regarding the memorial landscapes in Europe and in the Asia-Pacific realm have taken place in the past few decades as well as in recent years. In a regional review some of the most prominent changes and alterations at existing sites as well as new projects will be highlighted. First, the emphasis is on positive developments and trends. Eventually, major challenges to the existence of the sites and their improvements will be discussed.

Even at older existing memorial sites, like Dachau, new projects have been planned and implemented (see Hammermann 2021). Extensions of areas, which were off-limits like the former SS housing area (which was used by the Bavarian state police) or the area where the Dachau trials against the former SS guards and other Nazi officials were held (1946–48), are now in focus. Adding these areas to the Memorial Site increases it to the size of the concentration camp area during the late 1930s and in the 1940s. At that point, the Dachau concentration camp resembled a huge military-industrial complex, surpassing in size the old town and residential areas of the Town of Dachau. Also remarkable are the many more themes that are being highlighted in special exhibits or are shown in walks in and near the camp by local guides. The City of Dachau has taken on a greater sense of responsibility, with, for instance, locally organized events for January 27, 2024 (International Holocaust Remembrance Day; see info@news.kz-gedenkstaette-dachau. de). Collaboration between the Memorial Site and the City has become smooth and productive. Meanwhile research on special segments of life in and outside the camp has been conducted, with new results and interesting findings for the public.

One of the chosen themes is resistance inside and outside the Dachau Camp. It was at Dachau that Georg Elsner, initiator of a failed assassination attempt on Hitler in 1939, was shot on April 9, 1945, and his body was burnt at the crematorium on the same day. His motivation and life path were presented at a Dachau Memorial Site memory event. In addition, the Memorial Site offered a camp walk with emphasis on secret (forbidden) literary-biographical records of prisoners (see info@news.kz-gedenkstaette-dachau.de). These excerpts offer important inside views of the camp at Dachau. Similar practices occurred at the Kaufering satellite camps. At Kaufering I prisoners secretly kept a newsletter for distribution among fellow barrack mates.

DOI: 10.4324/9780367823795-15

A general trend is a greater willingness to document forms of resistance in Nazi Germany. Long seen as the actions of traitors (against the 'German Fatherland'), resistance groups are now widely recognized (Dowe 2021). An outstanding example is the members of the White Rose (*Weisse Rose*), students at the Munich University who produced critical flyers about the true situation in Germany in 1942/1943. The six core members of this group in resistance have become household names, in particular the siblings Hans and Sophie Scholl. They are remembered in multiple ways: at the places where they lived, where they took actions, where they were arrested, where they were put to death and where they are buried. Over a hundred high schools in Germany carry the names of Sophie and Hans Scholl. The entrance square to Munich University has been renamed in honor of the Scholl siblings; a permanent exhibit on the 'White Rose' inside the central university structure – where they once distributed their flyers – is well visited. Outside, a monument with copies of one of their flyers is found on the ground in commemoration of their courageous statements (Kronawitter 2014; Hartmann 2018, p. 487). The main place of commemoration for resistance groups and individuals is in Berlin, at the block where Claus Schenk Graf von Stauffenberg, a member of the German military, was shot after a failed assassination attempt on Hitler on July 20, 1944 (Hartmann 2018, pp. 486–487).

In recent years, the ('Aryan') women of the Rosen Strasse (*die Frauen der Rosenstrasse*) who quietly protested for the release of their Jewish husbands have found wide recognition. A little park with sculptures of the women and an exhibit at a nearby hotel remind the Berlin population of the events that played out in the Jewish section of town (Hartmann 2018, p. 487).

New centers and memorial sites came about in many places in Germany. In Frankfurt, a new Anne Frank Educational Centre, with an exhibit 'Anne Frank, a girl from Germany', has been established. The Frankfurt Jewish Museum added a Frank Family wing – in honor of the family who had resided in town since the 1600s.

A new approach regarding the memorial sites in Germany is a distinction between places in honor of the victims and places that are associated with the perpetrators ('Opfer' and 'Taeter' Orte). The latter type includes the Wannsee-Villa in Berlin where the SS decided on the 'solution of the Jewish question' on January 20, 1942, the Nuremberg Nazi Party Rally grounds and the second home of Hitler in the Bavarian Alps where fatal decisions were made during WWII. Most often, these sites have a documentation center which gives background information about the historical relevance of the places. In September 2023, a Obersalzberg Documentation Center, with the history of Hitler's Berghof, was opened to the public.

In Amsterdam, more members of the resistance are featured in the Verzets Resistance Museum, with special exhibits and new interactive IT. A trend in the design of new exhibits all over Europe is the digitally interactive mode which allows the visitor to address personal questions for further explanations (Hartmann 2017; Horstkemper 2021). This has also come into play at the new National Holocaust Museum in Amsterdam and at the Hollandsche Schouwburg memorial site, with multi-media shows. The exhibition technology at many memorial museums has been updated so that, in particular, young visitors can actively learn and participate.

In China, the exhibits at the Lu Gou Qiao Resistance Museum have been restructured several times to adjust to new research results about the war events in 1937–1945. An expanded and updated Nanjing Museum about the 300,000 victims of the massacre of December 1937/January 1938, with the exhibition assembly area, the memorial area, the Peace Park area and the collection area, is now visited by eight million people annually. Since December 13, 2014, a national memorial day has been created, with the memorial venue setting the stage for a ceremony held annually for the victims of the massacre. It is the 'dark tourism' experience at the very large site that has fostered the success of substantially increased visitation (Cui et al. 2020).

A memorial is also placed at Nanjing regarding the many Chinese women and girls who served as 'comfort women' during the times of occupation. Japanese historians have questioned the validity of the information as to the number of victims during the Nanjing Massacre and as to the enslavement of the women by the Japanese military (Tokushi 2022; Yamaguchi 2004)

In Germany, a major challenge to the now well-established memorial sites in honor of the victims of the national socialistic regime (1933–1945) is waged by right-wing populist party members, in particular of the Alternative fuer Deutschland (AfD) party. They gained wider acceptance in the states of former East Germany, with pre-existing anti-democratic traditions. The agenda of the AfD is to downplay what happened at the memorial sites and to disregard the major facts there. Members of these extreme factions have fostered a historical revisionism and broadened their appeal in the rest of Germany. AfD members have started to disrupt the tours and lectures at the prominent concentration camp memorials. Recently, the management of five memorial sites (Dachau, Bergen-Belsen, Buchenwald, Mittelbau-Dora and Neuengamme) decided to exclude members of the AfD from commemorative ceremonies at their sites (Arnold 2019; Die Welt 2023; Koelnische Rundschau 2023). In 2023/2024, allegiance with the AfD has increased consistently in the former East German states, up to about 30 percent according to polls, and to 15 to 20 percent in some of the older states of West Germany, whereas the traditional parties lost slightly or considerably in popularity. Still, the actions of the AfD are subject to examination by the Constitutional Protection Agency (Verfassungschutz) of the Federal Republic of Germany.

Similar developments and extreme political activities have been observed in other European countries. English language social media posts have become common outlets for the populist right-wing groups. Recent national elections in the Netherlands (November 2023) led to an upset, with far-right leader Geert Wilders and his Party for Freedom (PVV) garnering 37 seats, the highest number among all the Dutch political parties. What unites the PVV and AfD besides a populist approach? Their party programs are largely based on anti-immigration policies and European skepticism.

The situation in Japan has also seen the persistence of ultra-conservative movements honoring a clean and positive appreciation of the past. A 'war memory museum boom' (see Chapter 8) may have contributed to a wide range of information about the war events but right-wing Japanese politicians hold on to the ideals of a Greater East Asian War. New themes or more plentiful information about, for instance, the 'Comfort Women' issue have been largely rejected (as Tomoni Yamaguchi, a Japanese scientist based in the United States, observed in her detailed analyses of the right-wing scene: Yamaguchi 2020).

While the status of the memorial sites in Germany and their administrative and financial support system have been improving in recent years, negative or ambiguous perceptions of the missions of these sites persist in parts of the public. This is a troubling observation for the supporters of the memorial sites (still holding a majority in Germany's democracy). At the same time, it marks a continued challenge and task in the maintenance of the sites commemorating the events of the years 1933–1945.

References

Arnold, R. (2019). *KZ-Gedenkstaetten und die AfD – Der Kampf gegen die Verharmloser, Deutschlandfunk Kultur*, 6 August. https://www.deutschlandfunkkultur.de/kz-gedenkstaetten-und-die-afd-der-kampf-gegen-die-100.html.
Cui, R., Cheng, M., Xin, S., Hua, C., and Yao, Y. (2020). International Tourists' Dark Tourism Experience in China: The Case of the Memorial of the Victims of the Nanjing Massacre, *Current Issues in Tourism*, Vol. 23, Issue 12, 1493–1511.
Die Welt. (2023). KZ-Gedenkstaetten schliessen die AfD von Gedenkveranstaltungen aus, *Panorama*, 14 September. https://www.spiegel.de/panorama/afd-kz-gedenkstaetten-schliessen-rechtsaussenpartei-von-gedenkveranstaltungen-aus-a-b222ce28-4b43-444a-b5af-456c76302388.
Dowe, C. (2021). Widerstand ausstellen – Umrisse einer deutschen Erinnerungs-Topographie, *Geschichte in Wissenschaft und Unterricht*, Jahrgang 72, Heft ¾, 212–227.
Hammermann, G. (2021). Die KZ-Gedenkstaette Dachau – Zukunft der Erinnerung, *Geschichte in Wissenschaft und Unterricht*, Jahrgang 72, Heft ¾, 125–144.
Hartmann, R. (2017). Places with a Disconcerting Past: Issues and Trends in Holocaust Tourism, *EuropeNow*, Issue 10. http://www.europenowjournal.org/2017/09/05/places-with-a-disconcerting-past-issues-and-trends-in-holocaust-tourism/.
Hartmann, R. (2018). Tourism to Memorial Sites of the Holocaust, in P. Stone, R. Hartmann, T. Seaton, R. Sharpley and L. White (Eds.) *The Palgrave Handbook of Dark Tourism Studies*, London: Palgrave Macmillan, 469–507.
Horstkemper, G. (2021). Erinnerungsorte und Gedenkstaetten zur NS-Zeit im digitalen Raum, *Geschichte in Wissenschaft und Unterricht*, Jahrgang 72, Heft ¾, 228–230.
info@news.kz-gedenkstaette-dachau.de Zeitschichte Stadt Dachau Samstag, 27 January 2024. info@news.kz-gedenkstaette-dachau.de Biografisch-literarischer Rundgang, 24 February 2024.
Koelnische Rundschau. (2023). *KZ-Gedenkstaetten sprechen sich gegen Zusammenarbeit mit AfD aus*, 15 September. https://www.rundschau-online.de/politik/kz-gedenkstaetten-sprechen-sich-gegen-zusammenarbeit-mit-afd-aus-647208.
Kronawitter, H. (2014). Sophie Scholl – eine Ikone des Widerstands, in *Einsichten und Perspektiven*, Muenchen: Bayerische Landeszentrale fuer politische Bildungsarbeit, 14/2, pp. 80–91.
Tokushi, K. (2022). The Nanjing Massacre in Japanese Historiography and Education, in L. Wigger and M. Dirnberger (Eds.), *Remembrance – Responsibility – Reconciliation. Philosophische Perspektiven*, Berlin: J.B. Metzler.

Yamaguchi, T. (2004). *Japanese Right-Wing Responses to Comfort Women Studies*, Ph.D. Dissertation, Ann Arbor: University of Michigan.

Yamaguchi, T. (2020). The 'History Wars' and the Comfort Woman Issue: Revisionism and the Right-Wing in Contemporary Japan and the U.S., *The Asia-Pacific Journal*, Vol. 8, Issue 6.

Index

For Product Safety Concerns and Information please contact our EU
representative GPSR@taylorandfrancis.com
Taylor & Francis Verlag GmbH, Kaufingerstraße 24, 80331 München, Germany

* 9 7 8 1 0 3 2 8 5 2 1 5 7 *